Learner-Centered Teaching Activities for Environmental and Sustainability Studies

Loren B. Byrne

Editor

Learner-Centered Teaching Activities for Environmental and Sustainability Studies

 Springer

Editor
Loren B. Byrne
Roger Williams University
Bristol, RI, USA

ISBN 978-3-319-28541-2   ISBN 978-3-319-28543-6   (eBook)
DOI 10.1007/978-3-319-28543-6

Library of Congress Control Number: 2016933668

Springer Cham Heidelberg New York Dordrecht London

Printed on acid-free paper

Springer International Publishing AG Switzerland is part of Springer Science+Business Media (www.springer.com)

*For Kim and our frogs, fish, birds,
and gardens, and with gratitude for all my
teachers, mentors, and colleagues who have
inspired, nurtured, and supported my love
of learning and teaching.*

Preface

I don't recall exactly when I read Mary Ellen Weimer's book *Learner-Centered Teaching*; I may have still been in graduate school or my first post-Ph.D. year as a visiting professor. However, I'm certain that reading this book was a seminal moment that will forever influence my teaching philosophy. Weimer's premise resonated with me: "learner-centered teaching ... represents an entirely new way of thinking about teaching and learning tasks and responsibilities. It is transformational" (2002, p. xxi). As a young teacher, her arguments were revolutionary for me as I hadn't heard them before but found them to be immediately compelling and convincing. Now that I've had some experience putting the approach into practice, I can confidently echo her statement that learner-centered teaching is transformational. Overall, it is so dynamic and interactive that it can make lecturing feel dull and agonizing. Learner-centered thinking opens a door to more pedagogical creativity (as reflected by many of the activities in this book) that can lead to higher levels of student engagement, classroom energy, and teaching satisfaction. Quite simply, learner-centered teaching is more fun. Because of Weimer's book, my central career goal is to continually improve my learner-centered teaching practices.

Although I was a learner-centered convert in the first couple years of my current professor position, it took me a while longer to develop a mature and fully realized learner-centered philosophy. I quickly realized that I needed to adjust lectures to give students more responsibility for their own learning. In particular I remember asking one student why he wasn't taking notes during the lecture. He replied, very honestly, that he didn't have to because all the needed text was on the presentation slides which could be downloaded from the online course management system. This set off the nascent learner-centered teaching alarms in my head. I realized I was doing too much of the thinking for the students. As a result, I started to remove text from slides and insert blank spaces that I would fill in orally so at least they had to write the information down, or even better, I would help students contribute the missing information via Socratic dialogue.

Over time, I realized that such simple changes were not enough and that, to become more learner centered, I needed to reduce the classroom time spent by me talking about information on the slides. This commitment was especially strengthened by another classroom moment that I vividly remember, from three and four years ago. While lecturing, my mind was divided in two: one part was enthusiastically talking about the topic of the day in an attempt to impart knowledge on my students. Simultaneously, another part was observing the students more closely than usual. A few of the studious ones were paying attention and taking notes. But as my eyes scanned the room, I saw too many that were not mentally "there." A few were sleeping or clearly trying not to; others were daydreaming or just zoned out. Of course, the part of my mind that saw this immediately began to fret: *Why don't the students care? Why are they so irresponsible? Don't they want to learn?* Although this strong tendency to blame the students is common, with further reflection, it's easy to see that it's not wholly their fault. Who among us has not nodded off at inappropriate times, especially when sitting, listening passively, only to then suddenly wake up wondering when your nap began? I certainly have, including, for example, more times than I'd like to admit in my invertebrate biology class (at 8 AM—part of the problem for a night-owl like me). (Despite this, I went on to study invertebrate ecology in graduate school.) Surely we've all seen colleagues drift off to sleep during professional conference talks, departmental seminars, or committee meetings. It's an inevitable result of sitting passively without the responsibility for having to accomplish something in the moment—a biological outcome of a stressed and tired body seeking some shut-eye whenever possible.

Although perhaps inevitable and surely undesirable, seeing disengaged students during a lecture can be a dreaded, demoralizing—even enraging—classroom moment for teachers. In my particular instance described above, while one part of my mind was trying to concentrate on lecturing coherently, the other was asking, why are you doing this? What *is* the point? Aren't you partially responsible for the students' disengagement and opportunistic slumber? What should you be doing to reduce the probability of this happening? I left that class with a stronger commitment than ever to creating more active and engaging class meetings and reducing, as much as possible, the amount of time I spend talking at students in lecture mode. Yes, I still lecture; it can be a useful, if not necessary, tool in some cases, if just to vary the course's day-to-day flow. But one's teaching methods should be more diversified and engaging than lectures alone. With critical observations and reflections, it is apparent that lecturing sometimes—perhaps often—just doesn't work to promote excellent student engagement and learning, which should always be a teacher's primary goal.

One way to test the validity of this conclusion is to think about the most memorable learning and classroom moments from one's own time as a student. For me, I remember Dr. Hils asking volunteers (including me) to stand in the front of the classroom and create a human model of a cell membrane. He also had us create our own 3-D "life spirals" to illustrate the life history strategies of plants. In my general ecology course with Dr. Huehner, we sampled, identified, and counted beans to calculate biodiversity indices. In marine biology, he turned over one class meeting to each student so that we could be teacher-of-the-day (a smart approach that I use in some of my classes because having to teach someone else creates ideal motivation to spur one's own learning!). In addition to strengthening my knowledge about coral reefs, this experience helped solidify my desire to become an educator. Some of my other most enjoyable college experiences were in Professor Bourassa's photography classes; she made every studio critique session highly learner centered which forced us to refine our critical artistic thinking and communication skills. The common thread connecting these and many other of my classroom memories is that students were the center of the action—the ones doing the work. I suspect that readers can identify similar unique learning experiences from their own lives.

Learning experiences is a nice phrase to describe those memorable, participatory, probably enjoyable moments when students are awake, engaged, thinking critically, and doing something other than taking notes. What methods can teachers use to create more of these experiences? Alongside traditional approaches like laboratory experiments, research papers, and other projects, it seems that the number and diversity of innovative pedagogical tips and techniques has exploded in recent years. The creation of new terms and acronyms is hard to keep up with: PLTL, flipped classrooms, PBL, the 5 E's model, service learning, and POGIL, among others. In my mind, all of these fall into the umbrella category of learner-centered teaching. Although they are great general approaches that guide the creation of learning experiences, I have been frustrated in my teaching by the relative lack of concrete and immediately usable materials needed to implement learner-centered teaching in traditional classes that meet for 50–120 min in lecture-style classrooms (and specifically for the undergraduate level—many more resources exist for K-12 education). Throughout my career, I have developed many learner-centered materials and activities for such contexts but have found that few publishing outlets (that are of variable quality) exist for such college-level educational materials. Yet, I know that other professors have developed their own teaching activities and would also appreciate having other well-developed, easily-adoptable resources.

These two concerns—limited availability of teaching resources and publication outlets—are what prompted me to pursue publication of this book. My motivation was, in part, selfish: I wanted more teaching ideas for my own classes. However, the more important and valuable outcome of this book is sharing the great teaching activities of the contributing authors with the whole community who will surely benefit from their hard work and creativity. For those who already consider themselves to be learner-centered teachers, the resources will provide a wealth of new materials to use while also inspiring ideas for new activities. For those who are just beginning their teaching careers or haven't used many or any learner-centered teaching methods, I hope that the introductory chapter and collection of resources will motivate and enable them to become learner centered in their own educational philosophies and practices. No matter your current level and duration of learner-centered expertise and experience, I think it's fair to say that we can all strive to become better designers of valuable learning experiences for our students.

To be sure, becoming more learner centered requires significant time, patience, and commitment. I've experienced much self-doubt and anxiety along my journey, catalyzed by students who remained unengaged, still fail to do well, and complain about the "unfair" workloads on course evaluations. (Of course, some students have commented how much they enjoy and benefit from my teaching methods—but as many will empathize with, it's the negative comments that one tends to brood over while lying in bed!) However, on balance I feel confident that learner-centered pedagogy is simply the right way to approach contemporary education—a view which is supported by accumulating scientific evidence (as reviewed in Chap. 1). This is particularly true for environmental and sustainability studies which are among the most important areas for students to learn about given the world's many pressing socio-environmental challenges. The higher education community needs more published learning activities about these issues and many others. I hope this volume will be an inspiration and model for the development and dissemination of others. More immediately and personally, I hope the resources in this book help you accomplish your own learner-centered teaching goals.

Bristol, RI, USA   Loren B. Byrne
November 2015

Acknowledgements

This book would not be possible without the chapters' authors; I thank them for their time and dedication to creating and sharing their teaching activities and manuscripts. I am also grateful to all of them for serving as peer reviewers which made the editing process much easier. In addition, I thank my RWU colleagues Barbara Kenney and Charles Thomas for their review comments. Bruno Borsari, Rebecca Duncan, Kyle Emery, and Stephen Siperstein reviewed my introductory chapter and provided enormously helpful feedback that guided substantial improvements to it, which I greatly appreciate.

Many thanks to Sherestha Saini, from Springer, for her help in bringing this volume to fruition and answering my many questions. Roger Williams University provided me with the sabbatical necessary to undertake this time-intensive endeavor. I also thank the Dave Matthews Band, Fitz and the Tantrums, Jessie Ware, and all the other musicians who provided the essential soundtrack to my work. Last and most, I thank Kim for her love, support, patience, culinary skills, and companionship, all of which enabled me to complete this project.

Contents

Contributors

Carmen T. Agouridis Department of Biosystems and Agricultural Engineering, University of Kentucky, Lexington, KY, USA

Jessica Baum Department of Environmental Studies, Antioch University New England, Keene, NH, USA

Alice F. Besterman Department of Environmental Sciences, University of Virginia, Charlottesville, VA, USA

Bruno Borsari Department of Biology, Winona State University, Winona, MN, USA

Loren B. Byrne Department of Biology, Marine Biology & Environmental Science, Roger Williams University, Bristol, RI, USA

Sandra L. Cooke Department of Biology, High Point University, High Point, NC, USA

Erika Crispo Department of Biology, Pace University, New York, NY, USA

Patrick W. Crumrine Department of Geography and Environment, Rowan University, Glassboro, NJ, USA

Department of Biological Sciences, Rowan University, Glassboro, NJ, USA

Bethany B. Cutts Department of Natural Resources and Environmental Sciences, University of Illinois at Urbana-Champaign, Urbana, IL, USA

Kate J. Darby Huxley College of the Environment, Western Washington University, Bellingham, WA, USA

Rebecca Duncan English Department, Meredith College, Raleigh, NC, USA

Kyle A. Emery Department of Environmental Sciences, University of Virginia, Charlottesville, VA, USA

Jennifer Gaddis Department of Civil Society and Community Studies, University of Wisconsin-Madison, Madison, WI, USA

Jessica A. Gephart Department of Environmental Sciences, University of Virginia, Charlottesville, VA, USA

Curt Dawe Gervich Center for Earth and Environmental Science, SUNY Plattsburgh, Plattsburgh, NY, USA

Nathan Hensley Department of the Environment and Sustainability, Bowling Green State University, Bowling Green, OH, USA

Laura A. Hyatt Department of Biology and Sustainability Studies Program, Rider University, Lawrenceville, NJ, USA

Meghann Jarchow Sustainability Program, University of South Dakota, Vermillion, SD, USA

Matthew Kolan Rubenstein School of Environment and Natural Resources, University of Vermont, Burlington, VT, USA

J.J. LaFantasie Colorado State University Agriculture Experiment Station, Grand Junction, CO, USA

Department of Biological Sciences, Fort Hays State University, Hays, KS, USA

Kimberly F. Langmaid Sustainability Studies Program, Colorado Mountain College, Edwards, CO, USA

Walking Mountains Science Center, Avon, CO, USA

Timothy Lindstrom Nelson Institute for Environmental Studies, University of Wisconsin-Madison, Madison, WI, USA

Joshua Long Environmental Studies, Southwestern University, Georgetown, TX, USA

Jessa Madosky Biology Department, University of Tampa, Tampa, FL, USA

Bonnie McBain School of Environmental and Life Sciences, University of Newcastle, Callaghan, NSW, Australia

Corina McKendry Political Science Department and Environmental Program, Colorado College, Colorado Springs, CO, USA

Catherine Middlecamp Nelson Institute for Environmental Studies, University of Wisconsin-Madison, Madison, WI, USA

Erica A. Morin Department of History, Westfield State University, Westfield, MA, USA

Katharine A. Owens Politics and Government and Environmental Studies, University of Hartford, Hartford, CT, USA

Michael L. Pace Department of Environmental Sciences, University of Virginia, Charlottesville, VA, USA

Jae R. Pasari Science Department, Berkeley City College, Berkeley, CA, USA

Science Department, Pacific Collegiate School, Santa Cruz, CA, USA

Liam Phelan School of Environmental and Life Sciences, University of Newcastle, Callaghan, NSW, Australia

Zanvyl Krieger School of Arts and Sciences, Johns Hopkins University, Washington, DC, USA

Mercedes C. Quesada-Embid Sustainability Studies Department, Colorado Mountain College, Edwards, CO, USA

Courtney E. Quinn Sustainability Science, Furman University, Greenville, SC, USA

John E. Quinn Department of Biology, Furman University, Greenville, SC, USA

Emily S.J. Rauschert Department of Biological, Geological and Environmental Sciences, Cleveland State University, Cleveland, OH, USA

Nathan Ruhl Department of Biological Science, Rowan University, Glassboro, NJ, USA

Tyler M. Sanderson Department of Biosystems and Agricultural Engineering, University of Kentucky, Lexington, KY, USA

Heather E. Schneider Department of Ecology, Evolution, & Marine Biology, University of California, Santa Barbara, CA, USA

Andrew J. Schneller Environmental Studies Program, Skidmore College, Saratoga Springs, NY, USA

Stephen Siperstein Department of English, University of Oregon, Eugene, OR, USA

Marilyn Smith College of Business Administration, Winthrop University, Rock Hill, SC, USA

Christine M. Stracey Department of Biology, Guilford College, Greensboro, NC, USA

Lynn C. Sweet Center for Conservation Biology, University of California, Riverside, CA, USA

Seaton Tarrant Sustainability Studies, University of Florida, Gainesville, FL, USA

Rachel Thiet Department of Environmental Studies, Antioch University New England, Keene, NH, USA

Christine Vatovec Rubenstein School of Environment & Natural Resources, University of Vermont, Burlington, VT, USA

Madhavi Venkatesan Department of Economics, Bridgewater State University, Bridgewater, MA, USA

James D. Wagner Biology Program, Transylvania University, Lexington, KY, USA

Grace M. Wilkinson Department of Environmental Sciences, University of Virginia, Charlottesville, VA, USA

Joseph Witt Department of Philosophy and Religion, Mississippi State University, Starkville, MS, USA

Chapter 1
Learner-Centered Teaching for Environmental and Sustainability Studies

Loren B. Byrne

"Teaching succeeds when learning occurs."

(Davis and Arend 2013, p. 31)

"The one who does the work does the learning."

(Doyle 2011, p. 1)

Introduction

Change is often very slow in educational systems (as educators are—often pain-fully—aware). Nonetheless, the pace can sometimes quicken when a certain zeitgeist emerges, tipping points are passed, and critical masses of interest and action lead to big-scale, possibly dramatic transformations. This volume merges two aspects of education that have rapidly changed in recent years: environmental and sustainability education, and knowledge about effective pedagogy. Its central purpose is to disseminate engaging teaching activities that instructors can use to help increase students' environmental literacy: the knowledge, skills, and dispositions (values, attitudes, and motivation) for engaging in actions that effect positive environmental, sustainability and human well-being outcomes. More generally, this volume aims to inspire instructors to adopt learner-centered pedagogical practices that can improve student learning (see Box 1.1 for a note about the teaching-learning relationship). The target audience is primarily higher-education instructors (including graduate level), but high school teachers may find the resources appropriate for and adaptable to their courses, especially advanced ones.

Across all teaching levels, environmental and sustainability educators need diverse, well-developed, easy-to-adopt teaching resources to facilitate effective teaching that in turn fosters excellent learning (defined in this book to include

Electronic supplementary materials: The online version of this chapter (doi:10.1007/978-3-319-28543-6_1) contains supplementary material, which is available to authorized users.

L.B. Byrne (✉)
Department of Biology, Marine Biology & Environmental Science, Roger Williams University, Bristol, RI, USA
e-mail: lbyrne@rwu.edu

© Springer International Publishing Switzerland 2016
L.B. Byrne (ed.), *Learner-Centered Teaching Activities for Environmental and Sustainability Studies*, DOI 10.1007/978-3-319-28543-6_1

1

knowledge gains, skill development, and personal growth in affective dimensions). The unavailability of such resources can often be a limiting factor for those who wish to evolve their pedagogy and try new approaches, in large part because of time constraints on developing new materials. To be sure, many teaching resources exist, especially on the Internet where searches can yield varied ideas and lesson plans from diverse sources. However, it may not be easy (nor quick) to find such materials, and searches often end in frustration. Existing resources can be limited in many ways that preclude their utility. Some may be underdeveloped, with less-than-ideal rigor and sophistication for use in particular settings. They might be tailored to the creator's personal interests and thus have limited transferability for teaching more general themes. Some resources are deeply rooted in narrow disciplinary perspectives that preclude their use with students of diverse academic backgrounds and skill levels. Most often (in my experience), materials simply do not exist for particular topics and learning goals. These challenges (and others highlighted in the next section) suggest the need for additional initiatives to formalize and share the unpublished teaching activities that many instructors have developed for their own courses.

This book helps fill the teaching-activity resource gap for environmental and sustainability studies (especially for higher-education contexts) by presenting a diverse, highly polished collection of pedagogical materials. The practicality and usability of this collection arise from its emphasis on the daily practices of teaching. It takes a "user-centered" view by presenting concisely described teaching activities with explicit and sufficiently detailed guidance to enable their (relatively) quick and easy adoption. The volume is limited to activities that require minimal instructor preparation (with ready-to-go supplementary materials provided as possible), do not require elaborate setups or specialized materials, and can be used in many course contexts (e.g., small and large, introductory through advanced). All are designed for traditional (i.e., non-laboratory, indoor) classroom spaces, and most can be completed in one or two class sessions. No matter the implementation context, each activity can create a significant learning experience (sensu Fink 2003) in which higher classroom energy levels and student engagement lead to meaningful, long-lasting outcomes for students' personal lives, community connections, and professional pursuits.

The rest of this introductory chapter will further explore key issues that frame this collection of teaching activities. First, the larger context around and some pedagogical challenges of environmental and sustainability education will be discussed. Second, a brief review will be provided about the nature and implications of adopting learner-centered pedagogy to highlight the value of and need for the continued shift to this approach. To foster critical, honest reflection about the approach, a third section discusses some of its inherent challenges and trade-offs; the goal here is to help readers more thoughtfully reflect about how to implement it in a personally relevant, comfortable, and balanced way that fits one's personal context, which is necessary for its effective and sustained use. Finally, some concluding reflections will be given based on my work as editor of this volume and my own teaching experiences. Overall, I hope that the volume will inspire readers to use many of its exciting teaching activities to help their students become more environmentally literate citizens who help humanity work toward a more sustainable future.

Box 1.1. Teaching or Learning Activities?

Teaching and learning exist in a kind of mutualistic-symbiotic relationship; they are intricately linked (perhaps in an obligatory way) and ideally should benefit—or at least inform—each other through reciprocal feedback. In a learner-centered context, learning is the more important of the two because the purpose of teaching is to facilitate learning, as highlighted by the epitaphs that opened this chapter. Teaching activities should therefore always be designed as learning activities. Perhaps a preferable term might be "teaching-learning activities" to indicate the intimate coupling of the two processes that should be viewed as inseparable.

In this context, why was the phrase "teaching activities" used for the title of this volume instead of "learning activities"? Shouldn't the focus be on learning rather than teaching? Yes! However, since the volume's audience will be composed of teachers (this is an educator's resource, not a student workbook or textbook), readers will be examining and implementing the chapters from the perspective of the teaching role (e.g., preparation, guiding students). Second, the phrase learner centered is preferable to student centered because it emphasizes the learning aspect of being a student (as noted by Weimer 2002). Following from this, the title "learner-centered learning activities" would have sounded awkward due to its internal redundancy. In the end, the final title may not matter much, and these two phrases (learning activities and teaching activities) could be seen as interchangeable. The key insight is to remember that a book about teaching activities is simultaneously a book about learning activities, since learning must occur for teaching to occur (Weimer 2002; Davis and Arend 2013).

Environmental Issues, Sustainability Goals, and Educational Challenges

The topics of pedagogy and environmental/sustainability education coalesce around a central question: what teaching methods should educators use to help students help improve the world? In other words, how can educators help students learn how to live responsibly and sustainably on Earth to help make it—in all of its myriad environmental, human, and societal dimensions—a better, more sustainable, and thriving place for everyone, including other species in the biosphere's diverse ecological webs? It does not seem like an overstatement to suggest that—on a planet with limited, stretched-ever-more-thinly resources and a population of seven-billion-plus-and-counting people—few questions are as important as this one, especially in educational systems that seek to positively influence the future. As David Orr (2004) suggested, "… the worth of education must now be measured against the standards of human decency and survival—the issues looming so large before us in the twenty-first century. It is not education, but an education of a *certain kind*, that will save us" (p. 8, emphasis added). (In this context, it is important to clarify that this type of education is not, and should not be thought of as, a dogmatic, authoritarian approach that seeks to "brainwash" students into believing and acting in certain "correct" ways. Instead, the most effective environmental and sustainability education should seek to help students learn how to think critically, independently, and ethically about human-environment relationships, not specifically what to think and do.)

Orr's *certain kind* of education focuses on the realities of human-environment relationships in the twenty-first century. In particular, it is now unassailable that humans have changed the Earth system more over the past 60–100 years than at any time in our species' history, with dramatic, largely negative, consequences for biodiversity, the health and resilience of ecosystems, and therefore human well-being (McNeil 2000; MEA 2005). All students should understand the consequences of such changes in context of the natural and social scientific principles and evidence pertaining to humans' dependence on biodiversity, ecosystems, and favorable environmental conditions that give rise to ecosystem services that sustain human life, health, and happiness (Cardinale et al. 2012). Further, beyond content knowledge, sustainable societies need more citizens who have the willingness and wide range of skills to effect positive environmental and sociocultural changes. To these ends, educators have responded over the past several decades by developing integrated, transdisciplinary (inclusive of the natural and social sciences, humanities, and professional fields like architecture, engineering, and business) environmental and sustainability study courses, curricula, and degree programs (Vincent et al. 2013). Their emergence and growth advance Orr's vision for education that enables students to become sustainability leaders in their personal, public, and professional lives.

Concurrently, with increased environmental and sustainability education efforts, the past 20+ years have witnessed significant changes in scientific and philosophical understanding about the processes of teaching and learning (reviewed in the next section). In particular, evidence-based insights have led to numerous recommendations about how teaching activities and educational experiences should be designed (e.g., Bransford et al. 2000). One theme that has emerged is the necessity of using more engaging, hands-on, learner-centered pedagogies (instead of lecturing) to catalyze better learning (Kober 2015). Among the suggested techniques are project- and problem-based assignments that reflect "real-world" or "authentic" ways of helping students develop and apply their knowledge and skills. More generally, calls for "revolutionizing" the ways that educators think about and approach teaching and learning suggest the need to consider the pedagogical aspects of Orr's "certain kind of education"—that is, a kind of education that engages students in active, meaningful, and significant learning (e.g., Fink 2003). Such education is encompassed by the term "learner centered" (Weimer 2002).

Although the educational challenge of fostering significant learning applies to all disciplines and topics, it is particularly relevant to helping students develop their environmental literacy, a complex, multidimensional process and goal (McBride et al. 2013). If an accepted primary aim of environmental and sustainability education is for students to understand their own relationships to (including impacts on) the world and develop their capacities for effecting positive change, such education will have to help students more personally connect with the material as a means to enhancing their awareness, knowledge, and critical self-reflection abilities (e.g., Doyle 2011). Such education is needed to increase the chances that students will be prompted to modify (as needed) their affective and behavioral dimensions that relate to environmental and sustainability concerns. From this, a critical, if some-what obvious, insight emerges: achieving the laudable but challenging goals of environmental and sustainability education is a direct function of using effective, engaging pedagogies. Learner-centered methods are especially applicable to help-ing students process complex and challenging information from many disciplines; synthesize and apply knowledge to solve problems; consider diverse human per-spectives, cultures, and scenarios; and identify their own personal ethics, attitudes, and desired actions pertaining to human-environment relationships. Successful achievement of these outcomes requires that students be purposeful, reflective, and highly engaged participants in their own learning (Kober 2015).

The pursuit of more engaging learner-centered teaching approaches for environ-mental and sustainability studies can be challenging (as is true for any discipline). Many factors may limit or interfere with the ability (or willingness) of instructors to use more active pedagogical methods and materials. As discussed below, these include practical issues of the teaching context; a preference for focusing on content at the expense of explicitly helping students develop skills; and a reluctance (or uncertainty about how) to engage students' affective dimensions. (For the sake of brevity, the following discussion highlights generalized concerns without consider-ing nuances and exceptions. Although the issues will not apply to everyone and all situations, they are likely to be common across many teaching and institutional set-tings. See Kober (2015) for extended discussion about these and other factors.)

Practical Challenges

Many instructors, by choice or situation, teach in contexts that make the use of inno-vative learner-centered pedagogy more difficult (although not impossible). Courses may be designed to include a wide breadth of content, as with introductory surveys that serve as foundational prerequisites for subsequent classes. The perceived need to "cover material" as quickly and efficiently as possible (e.g., via lecturing) places pressure on instructors to ensure that students are "exposed to" information thought to be needed for future success (regardless of whether superficial exposure results in meaningful gains in student understanding, Kober 2015). This is especially true for environmental and sustainability studies which integrate an enormous breadth of content to create a more holistic, synthetic knowledge base. In addition, the goal of such content-centered courses may be to help students learn basic vocabulary, facts, and theories which may be seen as less relevant and amenable to innovative teaching (i.e., students should just be told what they need to memorize).

Further challenges arise from noncontent issues. Many classroom spaces remain traditionally configured with rows of desks and chairs (which may be immoveable) facing the lecturer. Many classes meet in short blocks of time (50–80 min) which can limit the use of project-/problem-based and other activities that require longer periods of engagement. Depending on institutional context, courses may have large enrollment numbers (e.g., >30 students) that frustrate the use of some pedagogies. (Much work, however, has focused on implementing active-learning techniques into large classes (e.g., Walker et al. 2008).) Further, although there is wide variation in instructors' professional responsibilities (including teaching loads) and locally available resources and opportunities, the materials, money, and/or time for finding, preparing, creating, and implementing new sophisticated and rigorous teaching activities are often limited. (As such, it is easier to do what has always been done than develop something new.)

Individually or in combination, these practical challenges mean that it may not be feasible for instructors to make all—or even a majority of—class meetings learner centered, hands-on, and innovative. Unfortunately, in many institutions these issues will probably remain for many instructors into the foreseeable future. To make pedagogical changes in such contexts, a key question must be addressed: how can learner-centered teaching practices and materials be integrated into traditional courses with lots of content, large enrollments, short meeting durations, inflexible classroom spaces, and instructors who are pressed for time and resources? Collections of scalable and easy-to-adopt resources geared toward such contexts, such as this one, are invaluable for assisting many instructors to integrate more innovative pedagogies into their traditional lecture-style courses.

Balancing Content and Skills

A second challenge for environmental and sustainability educators is addressing the tension between learning content and developing skills. Of course, these are not mutually exclusive, and most, if not all, instructors seek to help their students improve in both dimensions. Nonetheless, if for no other reason than limited time, a trade-off often exists in many teaching endeavors between these goals. In particular, this is due to the fact that rigorous skill development often requires explicit, purposeful, and iterative practice with focused, in-depth feedback. Although this makes helping students improve their skills more challenging (especially given the practical concerns described above), it is no longer—if it ever was—safe to assume that students' skills will automatically improve just by engaging in an activity once—or even multiple times—if instructional feedback is not provided. (This seems particularly relevant to reading and writing skills, which do not seem to develop with simple repetition for many students.) More explicit identification of skills-related goals and then aligning teaching practices with them are critical imperatives for education (e.g., as argued in the Vision and Change report about transforming undergraduate biology education; Brewer and Smith 2011).

For environmental and sustainability studies, skill development is often emphasized as part of their core *raison d'être*, especially in regard to systems and scientific thinking and problem solving (e.g., Wiek et al. 2011). Critical skills include gathering and evaluating information, evidence-based reasoning, applying and synthesizing information, and effectively communicating in many forms. As is surely obvious, teaching methods that actually engage students in doing these activities are critical for helping them progress (Bransford et al. 2000), especially through the use of many small, low-stake practice activities rather than one-time, high-stake, final projects. As such, instructors must be willing to "open up" time in their courses and curricula for multiple opportunities of sustained practice to improve students' skills. Of course, this is easier said than done in many contexts, given the many issues that compete for faculty's time, attention, and ability to provide students with thoughtful feedback. Further, sometimes teachers' love for content can outcompete the desire for helping students develop their skills. Although understandable (content remains important and is often fun to explore), Weimer (2002) maintains that, "When teaching is learner-centered, content is used, not covered ... to develop ... sophisticated skills necessary to sustain learning across a career and a lifetime" (p. xviii). In other words, knowledge acquisition and skill development should be integrated within holistic lessons. The tension between balancing content and skills might be resolved, in part, by the availability of compelling examples of pedagogy that focus intentionally and explicitly on skill development while also helping students learn basic information.

The Affective Domain

Another challenge that is particularly notable for environmental and sustainability education is engaging the "whole" student, including one's affective characteristics such as beliefs, values, and emotions, alongside motivation and optimism for "making a difference." Research has clearly shown that knowing "things" and having skills do not necessarily or directly translate into environmentally responsible attitudes and behaviors (e.g., Carmi et al. 2015; Sapiains et al. 2015). If a key goal of environmental and sustainability education is to prepare students to become agents of positive engagement and change, students have to be inspired to want to do this, beyond knowing why it's important and how to do it. Naturally, depending on their backgrounds and worldviews, many students who choose to take environmental and sustainability courses already have positive dispositions toward sustainability and have long been engaging in sustainable behaviors. Effective pedagogy can challenge them to clarify, refine, and develop their thinking and actions further. On the other hand, students who look negatively on (or actively refute) environmental and sustainability concerns can still be engaged in ways that (respectfully) ask them to reflect on, articulate, and justify their beliefs, values, and decisions. Rather than attempting to change their minds or "force" any views on them (as some misguided critics of environmental and sustainability education suggest is the goal), such engagement and outcomes are a crucial part of meaningful education that seek to promote the growth of students as individual, independent thinkers and help develop their reflective, metacognitive skills (which are valuable for reasons that go beyond the sustainability context). However, students' affective characteristics (emotions, values, attitudes) are often overlooked or purposefully shunned by instructors for a variety of reasons, including the difficulty of assessing or grading personal beliefs and outside-the-classroom, nonacademic behaviors (e.g., Shephard 2008). Nonetheless, pedagogy that asks students to reflect on their own views and consider alternative ones is essential for fostering significant and meaningful learning experiences (Fink 2003). It can also help students connect with and more deeply learn important content (Doyle 2011; Kober 2015). Suggestions for how to do this effectively and respectively, in open and supportive classrooms, may help instructors more readily consider and adopt exercises that integrate affective issues and reflective practices into environmental and sustainability courses with valuable outcomes for students and society.

These three challenges—negotiating the practical challenges of teaching, balancing content and skills development, and engaging students' affective dimensions—establish critical context for this book; providing resources that help overcome them is one of its main goals. (For additional references that describe environmental and sustainability study teaching activities, alongside some that provide generalized learner-centered teaching tips and techniques, see Electronic Supplementary Materials (ESM-A). The volume's scope has intentionally been constrained to include teaching activities that can be used in traditional college classes (meeting in short periods in rooms configured for lecturing) and that are relatively simple, inexpensive, and easy to implement (i.e., to overcome instructors' time limitations for preparation). Many of the activities (especially in Part 3) focus explicitly on helping students practice critical skills while also integrating valuable basic content; some can be used more than once with different topics to provide students with multiple opportunities to improve (e.g., Chaps. 9, 36, 37, 40). Some of the most unique and thought-provoking ones are those that directly focus on students' affective dimensions, especially through reflection about their personal views and behaviors (e.g., articulating personal statements of hope and purpose (Chap. 8), reflecting on personal tastes for food (Chap. 21), and considering empathy in the context of environmental justice (Chap. 29)). As a diverse and original collection of teaching resources, this volume will hopefully expand instructors' use of methods that can catalyze students' memorable A-HA! learning moments. More broadly, the significant learning about environmental and sustainability issues that can be fostered through the activities in this book will help advance David Orr's "kind of education" that is needed now and well into the future.

A Shifting Paradigm Toward Learner-Centered Pedagogy

David Orr is not the only person who has called on educators to reconsider how they approach teaching. In a seminal article, Barr and Tagg (1995) emphasized the need to shift the focus of our educational paradigm from teaching to learning. They argued that the dominant "teaching paradigm" (what Davis and Arend (2013) call the "lecture paradigm") is a teacher providing information to students (also see Fink 2003). In this context, Weimer (2002) noted that, historically, "most (educators) assumed that learning was an automatic, inevitable outcome of good teaching and so we focused on developing our teaching skills" (p. xi-xii), at the expense of considering whether the teaching was enhancing learning. A key limitation of this perspective, as she suggested, is that "the learning outcomes of teaching cannot be assumed or taken for granted" (Weimer 2002, p. xi–xii). This is especially true for lectures, which—while having some appropriate and necessary uses (see below)—are limited in many ways, especially their inability to foster the deeper, significant learning and sophisticated skills that most educators desire and students need (Ambrose et al. 2010; Davis and Arend 2013). This inability stems from a key concern that is intrinsic to all human learners: our naturally limited attention spans. Surely, all lecturers (including seminar speakers) have seen audience members (not just students!) nodding off or losing focus in the middle of a talk. (Even punctuating lectures with simple moments that engage the audience directly, like a question, can (re)capture attention.) Such limitations are revealed by research that has helped propel the shift toward a "learning paradigm" as called for by Barr and Tagg (1995).

Over the past two decades, the calls for pedagogical change have been increasing in number and urgency, catalyzed by improved scientific understanding about the nature of learning (Ambrose et al. 2010; Kober 2015). Insights from research help answer Weimer's (2002) crucial question: "What do we know about learning that implicates teaching?" (p. xii). We now know that "learners need to be actively engaged in making meanings of new information and assimilating it into their own contexts" (Davis and Arend 2013, p. 9). We know that "learning skills as sophisticated as those needed by autonomous self-regulated learners do not develop simply through exposure to the content of disciplines. They must be taught …." (Weimer 2002, p. 16). In other words, we know that knowledge and skills cannot just be given to students; they must be developed by each student individually; this is the constructivist view of learning (Bransford et al. 2000; Kober 2015). More than ever, we also know (based on quantitative evidence) that a diversity of pedagogical methods that actively engage students promote their learning and overall success better than lectures alone (e.g., Freeman et al. 2014; Andrews and Frey 2015; Jensen et al. 2015).

In sum, the core conclusion from the still-evolving, but well-established, "science of successful learning" (sensu Brown et al. 2014) is clear: learning is enhanced when learners participate in engaging activities rather than passively listen and take notes during a lecture. Active learning helps students build stronger mental models to organize knowledge and facilitate its future application (Ambrose et al. 2010; Kober 2015). As presaged by Barr and Tagg (1995) in their description of a "learning paradigm," the purpose of educational systems "is not to transfer knowledge but to create environments and experiences that bring students to discover and construct knowledge for themselves, to make students members of communities of learners" who are collaborating to help each other learn (p. 15). In other words, as emphasized by Bransford et al. (2000), "schools and classrooms must become learner-centered" (p. 23) (a phrase also used by Barr and Tagg (1995) and brought further into the spotlight by Weimer (2002)).

The shift to learner-centered pedagogy brings a set of crucial implications clearly into focus. First, as introduced above, it suggests the need to consider the constructivist view of learning, which emphasizes that each person must identify and internalize all the "pieces" of information needed to build her own "puzzle" of understanding, make sense of it, and connect it to what is already known (Bransford et al. 2000; Kober 2015). Since the process is, at a fundamental level, neurological—learning literally requires making neurons in the brain fire and rewire to create new connections—teachers, ultimately, cannot force learning to happen in others (Brown et al. 2014). For learning to occur, students must decide that they want to learn and intentionally activate their brain cells (Handelsmen et al. 2007). Although instructors can help initiate this, learning is dependent on the learner's willingness to purposefully engage (Blumberg 2009). In other words, "learning is not something done to students, but rather something students themselves do" (Ambrose et al. 2010, p. 3). From this constructivist view, the primary (and ultimate) responsibility for learning shifts to the student, giving deeper meaning to the phrase "learner centered"—the student's brain is literally the center of the learning process (Weimer 2002).

Following from this constructivist view of learning, an additional implication is that students emerge more clearly as unique individuals (rather than blending together into a collective class) who have varied backgrounds, perspectives, and learning needs. For pedagogical practice, this insight raises several key issues relating to personal reflection and metacognition (which relates to engaging students' affective domain as introduced in the previous section). First, it may be helpful, if not necessary, to have students retrieve and explore their preexisting knowledge (Kober 2015). This can both remind them of what they have already learned (which serves as a basis for connecting and organizing new information) and expose gaps, possible biases, and misconceptions that may need to be changed so as to not interfere with future learning (Bransford et al. 2000; Ambrose et al. 2010; Barkley 2010). In a more immediate time frame, asking students to reflect on their own learning and understanding (e.g., what did you learn? what is its significance? what are you unsure about?) can reinforce content and skills via retrieval practice and help them and instructors identify confusion and issues that need more attention to improve learning gains (Brown et al. 2014, Kober 2015). Third, inviting students to reflect on their own personal opinions and feelings (relating to any course issues including content and process, personal and societal relevance, and students' dispositions) is important for helping them connect with course content, thus promoting more significant learning gains (Fink 2003; Doyle 2011; Kober 2015). Further, asking students to share and discuss these reflections with their peers creates powerful learning moments that are truly learner centered (while helping them recognize the diversity of experiences and views in a group). Fink (2003) suggests that students "need to spend time reflecting on the meaning of the experiences and new ideas they acquire" if they are to become more adept at internalizing and making sense of information (p. 106). Such explicit metacognition is an important part of "helping people take control of their own learning" (Bransford et al. 2000, p. 18), which should be a primary aim of learner-centered education. Overall then, the inclusion of reflective practice—about previous and new knowledge, feelings, and opinions—is important and valuable for creating truly learner-centered classrooms.

A third implication of the learner-centered philosophy is that the roles of the instructor must be reconsidered. In context of the expectation that students are individuals who must construct their own knowledge and take more control of their own learning, educators become more responsible for instructing students what they need *to do* to learn, rather than strictly what to know. Weimer (2002) suggests that learner-centered teachers "position themselves alongside the learner and keep the attention, focus and spotlight aimed at and on the learning process" (p. 76). As such, they are guides of learning, helping students work through learning activities (Fink 2003; Blumberg 2009). Some teachers may feel uncomfortable with this shifting view if they perceive that their role is as a "content expert" who passes information on to students. However, as Doyle (2011) emphasized, "the role of teacher as expert does not change when moving from a teller of knowledge to a facilitator of learning. What changes is how this expertise is used" (p. 53).

The expertise of the instructor in a learner-centered classroom is applied to the design and implementation of learning activities that catalyze students' learning (Weimer 2002; Blumberg 2009; Kober 2015). That is, instructors create and manage the situations in which students will do the work necessary to learn (Barkley 2010; Weston 2015). From the teacher's perspective, this responsibility is naturally teaching centered: what do I, as the teacher, have to do? In this framing (which is needed at a practical level), it may be easy to lose sight of the students and their learning. Thus, as posed by Doyle (2011), the key learner-centered question for the teacher's work is, "how will (my) instructional decisions optimize the opportunity for students to learn the skills and content of the course?" (p. 2). As reflected in this question, the value of the learner-centered paradigm is that it continually—even forcefully—reminds educators to refocus on the ultimate goal of teaching: students' learning (Box 1.1). Nonetheless, instructors retain critically important roles and responsibilities with this paradigm: they can and must still talk, sharing their expert knowledge and providing essential feedback. The key focus is for instructors to evaluate their activities in context of learning-focused questions: how will my decisions and actions facilitate students' learning? What can I ask students to do to discover information, construct meaning, and practice skills? What questions and guidance can I provide to help them fire their own neurons and become more aware of their own thinking? In a learner-centered classroom, such questions are answered in ways that emphasize students', instead of teachers', responsibilities for doing more of the work required for learning.

Thinking of teaching as the process of designing "student-learning experiences" emphasizes another implication of the learner-centered view: a need for instructors to be more reflective and intentional about what they and the students are doing and why, in part by identifying clear, explicit goals (Kober 2015). (This contrasts with the "teaching-paradigm" view in which relatively less planning is needed to create lectures and exams, and no consideration about what teachers and students will do is needed; teachers lecture, and students take notes and tests.) This need brings critical questions more clearly into focus: what is most important for students to know and why? What do they need to do to learn the chosen topics? What can students already do that will facilitate their engagement? What activities will best accomplish the learning goals? How can students show that they have achieved the goals? How can expectations be communicated to students more clearly?

Although specific answers to these questions are as varied as the courses in a college's catalog, a general framework that can guide instructors' reflection about and development of learner-centered courses and individual class meetings is that of "backward design," described by Wiggins and McTighe (2005). This "results-focused" (Wiggins and McTighe 2005, p. 15) approach is learner centered, in part, because it emphasizes, as a first step of designing learning experiences, the need to identify what students should be able to do after an activity is completed, i.e., the desired learning outcomes (also see Weimer 2002; Driscoll and Wood 2007; Doyle 2011; Davis and Arend 2013). Well-articulated outcomes should be written with precise verbs (e.g., explain, synthesize, create) to indicate the expected student actions that will demonstrate their learning and skill development (Kober 2015). (Such learner-centered outcomes differ from course- and content-centered ones often phrased as "this course will explore ….") Even for basic knowledge gains (i.e., students should come to know something), outcomes should still be stated with active verbs (e.g., list, state, describe) rather than the vague construction of "students will understand" which does not indicate precise learner-centered abilities.

Well-articulated learning outcomes serves several valuable purposes in a learner-centered classroom. For instructors, they are guides within the backward design approach for choosing and implementing teaching activities, including project assignments and exam questions. In this sense, outcomes help keep instructors "honest" about what they really intend for students to learn and do and serve as reminders to avoid extraneous, tangential topics and assignments that would not directly help students meet the outcomes (Kober 2015). (See Wiggins and McTighe (2005) for a thorough overview of backward design which is outside the scope of this discussion.) When outcomes are shared with students, the verbs clearly communicate what is expected of them, thus focusing their attention and fostering motivation (i.e., this is what I need to work toward). For both students and instructors, learning outcomes serve as guide posts to evaluate progress toward desired results. With clearly stated expectations in mind, instructors can focus on providing students with more directed feedback. In turn, students have more context for interpreting feedback. For instance, outcomes can be referred to when discussing why students received a certain grade or did not receive credit for an exam answer or part of an assignment (i.e., "you failed to meet this outcome; you could not satisfactorily do this"). Overall, developing and using learning outcomes are important within a learner-centered approach because they help keep instructors and students goal oriented and focused on desired learning (Kober 2015). Their ultimate value was noted by Doyle (2011): "You cannot know if your time (teaching) was well spent if you don't know what you wanted the students to learn" (p. 53).

A final implication of learner-centered pedagogy is its ability to change the dynamics and energy levels of classrooms. When implemented, learner-centered teaching activities often replace the one voice of the lecturer with the many voices of students. Rather than quietly taking notes, students are talking to each other while solving problems, debating issues, sharing reflections, synthesizing information, and teaching each other. They are more likely to be interested, attentive, and motivated (or at least focused on achieving an immediate goal such as completing a worksheet). With certain activities, they might even be having fun, unaware that they are learning. While students work, the instructor walks around the room, talking with individual students and groups to answer questions, provide guidance, and challenge them to think more deeply. Such interactions can lead to serendipitous moments of eye-opening learning (for students and instructors) that make class meetings gratifying—the kinds of informal but excellent learning that may not be catalyzed by lectures. For the instructor, such dynamics can reveal more about students' thinking than can be observed from the front of the classroom while giving a lecture. In sum, adoption of learner-centered pedagogy can increase the levels of inspiration, engagement, and excitement in a classroom, all of which make the educational experience more rewarding, enjoyable, and memorable for students and teachers alike.

Challenges and Trade-Offs of Learner-Centered Teaching

A primary goal of this volume is to provide resources that help instructors overcome some of the challenges (highlighted above) of adopting learner-centered pedagogy. However, having a collection of teaching activities is not a panacea for all the implications that accompany adoption of learner-centered educational methods. These implications include inherent challenges and trade-offs that are not so easily overcome—in fact, using more teaching activities can increase these concerns. Most simply, sometimes activities just don't work as well as envisioned and hoped, despite the best design and implementation efforts. In addition, three others will be discussed briefly in this section: student participation and resistance, balancing the "content-class time" trade-off, and managing a (potentially) larger instructor workload. (See Weimer (2002), Blumberg (2009) and Kober (2015) for additional discussion of these and other concerns.) Highlighting such challenges and trade-offs is not intended to frustrate or deter readers from the learner-centered approach; rather, acknowledging them as legitimate and serious concerns should help readers consider and implement the approach more thoughtfully and successfully. Critical, thorough, and honest consideration of challenges and trade-offs is needed to help prevent cynicism, burnout, and abandonment of the approach. Although the challenges of learner-centered education may cause teachers and students alike to, at times, clamor for the relative simplicity, familiarity, and ease of lecturing and notetaking (which may in some instances be justified, see next section), I maintain that the challenges are not so severe that they provide just cause for disparaging the overall positive aspects of learner-centered approaches.

Student Participation and Resistance

Although active-learning pedagogy can increase the probability of student success, their use cannot guarantee that all students will participate fully and/or achieve learning outcomes. An activity that engages some students may bore or stupefy others (Weimer 2002). (Few things are as frustrating as seeing a great activity fall flat because a particular group of students don't do their part—especially when it has worked wonderfully with others.) A key challenge is helping students understand and accept—perhaps even support—the rationale for learner-centered pedagogy (Kober 2015). Clear communication about the rationale and goals for instructional decisions and activities may help achieve student buy-in (Weimer 2002; Blumberg 2009). Further, students should be continually reminded that the "responsibility to learn is theirs and theirs alone" (Weimer 2002, p. 103–104) and that they must be critically engaged to achieve the learning goals. For example, on the first day of my classes, I remind students that learning is, at a basic biological level, a process of making synapses among neurons in the brain fire; because I cannot enter their brains to make this happen, they are the only ones who can fire their own synapses. My responsibility is to provide some of the "fuel for the fire" that will help them do this, but neither I nor anyone can *make* them learn; they must *want* to learn and be motivated to take the appropriate steps to do so. Such introductory framing must be followed up throughout a course with pedagogical decisions that reflect this rather than counter it. Otherwise, students may become comfortably, passively numb (perhaps even cynical) and be resistant to an occasional learner-centered activity. Creating a truly learner-centered environment that overcomes initial resistance and frustration takes commitment, patience, and time from both instructors and students (Kober 2015).

Even with provided context and sustained use of learner-centered pedagogy, some students may remain resistant to doing more work in the classroom (especially if their other courses are lecture dominated). This is particularly true for students who have never been expected to take a more active role in their own learning or don't take classroom activities seriously, viewing them as "busy work" that wastes time (Walker et al. 2008). (This challenge may decrease as more instructors, in high schools and higher education, use effective, well-designed active-learning lesson plans.) A tough fact for teachers to accept is that it is not realistic to expect that all students will be excited and motivated by, and successfully learn with, all learner-centered teaching activities for various reasons, many of which are likely to be outside the teacher's control. For example, in my introductory ecology and evolution course, I try to use at least one active-learning technique in every class meeting (including ones that fill the entire session). Although these have engaged most students, a few have not participated fully, either doing the bare minimum needed or nothing at all (including by missing class). In an extreme case, I confronted one such student, noting that he did not do well on the previous exam and was now not engaged in the learning activity. I don't recall his exact response (he indicated that he was a philosophy major), but it was clear that he was not amused by the course's expectations for his active participation; after this, he did not attend class for the rest of the semester and ultimately failed because he did not withdraw. When faced with such cases, we should not let ourselves be haunted by failing to meet the "perfect standard" of engaging every single student in every activity; this is an unrealistic goal because we never will. In this context, key challenges are to remain focused on the positive aspects of our pedagogical choices; give attention to students who are engaged, motivated, and learning; and adjust activities as possible to increase the chances that they will be even more successful with more students the next time.

The Content-Class Time Trade-off

Although active-learning pedagogy can improve learning (e.g., Freeman et al. 2014), implementing it generally requires more class time than just having students take notes from lectures (which is attractive, in part, for its temporal efficiency of "covering" a lot of content even if its learning efficiency is not high—in one ear and out the other, as the cliché goes). As teachers know from their own education, constructing one's own knowledge is a time-intensive process! Thus, the decision to allocate class time to activities, even relatively short ones, creates a challenging trade-off: if it takes longer for students to learn one topic via an activity, less time is available for other topics. The implication of this is that learner-centered class meetings may include fewer content details in exchange for using activities that help students more deeply explore the most important concepts and overarching, organizing principles (which may ultimately be more important anyway for supporting course- and program-level outcomes, e.g., see Brewer and Smith 2011 for biology examples). A challenge for learner-centered teachers is becoming comfortable with this trade-off and thinking more critically about what content and learning outcomes individual class meetings (and a whole course) should include and exclude (as emphasized by Wiggins and McTighe's (2005) emphasis on "big ideas" and "essential questions"; also see Kober 2015).

Does this trade-off imply that, if learning activities are used, students will inevitably miss important basic content and that learner-centered classes will be less rigorous? No, not necessarily. Learner-centered instructors must emphasize more emphatically to students that they are responsible for acquiring important information outside of class through readings, videos, and other projects (an approach now commonly known as the "flipped classroom," e.g., see Jensen et al. 2015 and references therein). (Of course, basic content can be introduced to students via in-class activities as well.) Providing students with learning outcome statements can help focus them on what they need to know. In addition, to ensure their engagement, students may need strong "encouragement" to engage outside of class through required, graded assignments that will hold them accountable (e.g., worksheets, written summaries, or quizzes). Subsequently in class, the information can be reviewed, clarified, and, more importantly, applied, synthesized, and evaluated as part of learning activities. This facilitates the shift of the instructor's role from "teller of information" to "helper of information use and processing" and, in the context of assessing the work students do outside of class, "checker and enforcer of independent learning." Although students may complain about or resist being "forced" to do such work outside of class ("this is in-class lecture stuff!"), such an expectation follows naturally from the learner-centered approach because they are given more responsibility for taking control of their learning. Of course, instructors can also be frustrated by unmet expectations and lackluster student performance. As noted elsewhere, the key challenge is to feel confident about the instructional approach and clearly communicate to students that the pedagogical choices have been made to increase the potential for their best learning.

Instructor's Workload

Just as the out-of-class responsibilities for students become a more central focus in a learner-centered course, so does the instructor's preparatory work. The challenge here is that the workload is likely to increase as compared to using a traditional lecture-dominated mode of pedagogy. Developing and implementing new activities and assignments that are critically aligned with learning outcomes—especially over a whole course—is not easy, as most teachers know well. (Recognizing that various ways exist to implement learner-centered teaching, the comments here pertain primarily to the use of in-class learning activities that often—but not always—have students do work that creates tangible products such as answered worksheets or short writings.) In my experience, the challenges of increased time commitment arise from three main aspects of learner-centered pedagogy; although these may not universally apply to all instructors and activities, they exemplify trade-offs that may impact choices about how—and how often—to utilize teaching activities and associated assignments.

First, in its full implementation, the learner-centered approach emphasizes the identification of student-learning outcomes (see above). Choosing and articulating them for individual lessons and class meetings can be time-consuming because of the focused attention and critical reflection needed for this work: answering the key questions, "what is it, really, that I want students to be able to do; why this and not that?", is not always straightforward. In my experience, the process becomes much easier with practice, and once outcomes are identified for a course, it take much less time to revisit and revise them in the future. Nonetheless, I have often been too busy, distracted, or—of course it happens—mentally tired to write learning outcomes for a particular class session or activity. Sometimes, I can sense my own confusion about the big-picture purpose of an activity or class session when explicit outcomes are not prepared; this reflects their value for clarifying teaching and learning aims. (As such, explicit outcomes are provided for each chapter's activity in this volume.) Although it would be ideal, expecting to always have outcomes to guide every moment of teaching and student learning is not practical. However, this trade-off should not limit the use of engaging activities that can still foster learning even if they lack associated outcome statements.

A second key challenge for learner-centered instructors' workloads is the time needed to prepare instructional materials (including those to guide students' out-of-the-classroom learning). In my experience, this is likely to be the biggest hindrance to moving away from lecturing because it is very difficult to find or create new activities (especially high-quality ones that are aligned with outcomes) without substantial preparation—the time for which is often not available in the few days (or—often for many instructors, including me!—the night) before a class meeting. The trade-off here is to simply accept that a lecture may have to suffice in such instances (but one that, hopefully, integrates at least some interactive moments to make it more engaging; see Kober (2015) and the general resources in Electronic Supplementary Materials-A for some tips and techniques). To balance this trade-off, instructors could strive to choose a few class sessions in each course for which they begin planning learning activities weeks in advance. This may be easier said than done, but such efforts are needed to begin, and then iteratively continue, transforming courses to make them more dynamic (Blumberg 2009). On the upside, after an activity is used, it will be readily available for future use. In addition, more work is needed in the higher-education community to formalize and disseminate teaching resources (and create more peer-reviewed publication outlets for them). It is hoped that this volume will inspire such efforts which are invaluable for helping instructors overcome the frustrating prep-time challenge.

Finally, one key challenge that—unfortunately—cannot be so easily alleviated by past work and increased resource availability is the management of student-generated products that often accompany learner-centered activities (worksheets, minute papers, etc.). This includes collecting, reviewing, and evaluating them, scoring them as needed, and, as necessary and possible, providing students with feedback to guide their learning. For me, this is the most time-consuming, and sometimes tedious, aspect of taking the learner-centered philosophy to heart and implementing it as fully as possible (and I work at a smaller institution with most class sizes of 10–30; this challenge certainly scales up proportionally with increasing roster lengths). If at least one learning activity is used for each class period (as I strive to do, in-class or assigned as preparatory homework), a lot of student products are generated quickly. In addition to using them as formative assessment to help keep my finger on the pulse of student learning, I also use these products to award "participation points" (similar to attendance points but requiring that the students do something to show their engagement). Possible ways to alleviate this aspect of the workload, among others, are to not collect or score the student products, only review a sample of them, or simply check them as pass/no credit. More so than other challenges, decisions about how to manage the trade-off of student products with instructor workload are certainly an individual decision that depends on an instructor's particular preferences and context; a magic formula doesn't exist, and it may be necessary in some contexts to use learning activities much less than every class meeting to reduce this workload challenge.

Being honest and reflective about the challenges and trade-offs described above is a necessary step toward implementing an effective learner-centered teaching approach. Indeed, for each individual instructor, the biggest challenge may be the ongoing work of finding an individualized approach that "works"—that comfortably resolves the many challenges in context of all the other demands on one's time. I wish I could provide lots of specific, detailed advice about how to respond to the challenges described above, but I don't think that's remotely possible. I haven't found solutions that are applicable across all my own courses and semesters (which vary widely), much less the unimaginable range of others' situations. However, one critical piece of advice that is often emphasized, and I emphatically agree with, is to work slowly, incrementally, and iteratively rather than to try to transform all of a course's class meetings to be learner centered all at once (Weimer 2002; Blumberg 2009). (Also see Chap. 6 in Kober (2015) for some general suggestions about responding to challenges.) Another key recommendation I can provide is to talk to colleagues about the day-to-day struggles of teaching; others can provide helpful advice, but if nothing else, it's reassuring to discover that many of us face the same concerns and difficult choices. For instance, despite my commitment to the learner-centered philosophy, I sometimes struggle mightily to find a good balance between implementing it strongly and maintaining a manageable workload. On some (many?) days, I daydream of assigning less work to students so that I would also have less to do. As I have been trying my best to create more and more learner-centered lessons in my courses, I often have to intentionally remind myself to remain committed to my core pedagogical beliefs while not feeling guilty when choices are sometimes made that reduce the overall learner centeredness of my teaching. As such, my best advice to others with similar concerns might be to not give up on learner-centered teaching even if it its associated challenges feel overwhelming and trade-offs must be made that ultimately reduce the degree to which it can be adopted. In other words, it may be, in many instances, an honest and practical reality that "100 % learner-centered perfection" is an impossible dream. Of course, this doesn't mean that it is not a wonderful dream or that continual attempts to achieve it should be avoided. Even adding one more engaging learner-centered activity to a course might make a positive difference for one student's learning.

Concluding Comments and Reflections

Learner-Centered Lecturing?

Taken to its extreme, the learner-centered philosophy suggests that lecturing (defined here strictly as continuous, one-way speaking by the instructor) should be completely abandoned (as many have argued in countless articles and online blogs). Is this necessarily true, even in the strongest of learner-centered classrooms? In my opinion, no—that's an unhelpful suggestion. Lectures should be retained as part of learner-centered pedagogy if for no other reason than to add variety to the classroom experience (active learning in every single class could be just as monotonous as 100 % lecturing). Sometimes, a well-designed lecture can engage students very well (e.g., through captivating storytelling), focus their attention on key issues, convey enthusiasm for the subject, and illustrate (model) how a professional thinks about and organizes information (Bland et al. 2007; Kober 2015). For practical reasons, lectures may sometimes be needed when, for whatever reason, an alternative pedagogy isn't available for a particular topic. Further, many ways to engage with the world outside of classrooms expect people to have critical listening and note-taking skills (legitimate outcomes in their own right) that can be practiced during classroom lectures (Worthen 2015). For instance, many professionals attend seminars at conferences and work, where they are expected to pay attention to and learn from a speaker. Even listening to the news on TV, the radio, or a podcast exemplifies "being talked at" as a way to gain information. Given this context, a key challenge for learner-centered educators is helping students learn how to learn better when listening to speakers, no matter how boring and ineffective it may seem (MacKeachie and Svinicki 2006).

A better (and more realistic) question about lecturing is when and how the technique can be used more effectively in a learner-centered context (e.g., deWinstanley and Bjork 2002; Bland et al. 2007; Malik and Malik 2012; Gregory 2013). Most likely, a learner-centered teacher will lecture more sporadically and for shorter durations, using presentations to frame, punctuate, and complement the flow of other activities, including to provide critical "just-in-time" introductory information and instructions (as is suggested by some of this volume's chapters that provide example introductory presentations in the supplementary materials). At a minimum, pure seminar-style lectures can be made more interactive by embedding Socratic dialogue and short learner-centered activities within them (e.g., clicker or discussion questions, think-pair-share moments; Gregory 2013; Kober 2015; see Electronic Supplementary Materials-A for general resources that describe such methods). (Such mixed-method lessons might still be referred to as "lectures" reflecting the varied ways the term is used; see Hora 2014). A key consideration is to identify when a straight lecture by itself will be insufficient to help students develop the critical skills identified in learning outcomes; the student-centered verbs of outcomes can be used to develop activities that interrupt the instructor's talking and provide students with moments to do something other than listening and taking notes (e.g., explain, reflect, synthesize, calculate; Kober 2015). Such simple adjustments to facilitate students' active processing of information and construction of their own understanding can go a long way toward shifting a classroom to be less lecture dominated and more learner centered.

Content of the Volume's Chapters

Teaching activities can be presented in a variety of ways, ranging from very formal and detailed to informal and brief; certainly, one "right" way does not exist. In some contexts (especially K-12 education), activities may have to be prepared using standard "lesson plan" templates; other educators may use formats introduced to them in their formal training. For publishing, journals and websites vary in their requirements. When developing this book project, I chose to start with an empty page and think about what I, as a practicing teacher, would want included in the description of a teaching activity. This reflection guided my decisions about the standardized outline for each chapter (with one activity per chapter; an overview of each chapter's structure is provided in Box 1.2). Without provided guidelines, contributing authors would surely have developed their chapters in widely divergent ways. Thus, to ensure coherence within the volume, all chapters are identically structured and have been edited for consistency in tone and style. Further, I felt that brevity should be a key point of editorial emphasis; this allowed more activities to be included in the volume and should increase their attractiveness (in terms of readability and ease of adoption) and probability of more widespread use. Trade-offs surely exist for these two decisions, and some readers may be left wanting more or different information from each chapter. However, the information included in each chapter should provide the minimum information needed to make each activity "ready to go, right off the shelf."

Assessment of Learning Outcomes

Assessment of student learning (i.e., evaluating student work to provide comments, assign grades, or for other purposes) is a critical part of teaching, especially in a learner-centered classroom, where feedback is viewed as essential for improving learning (Weimer 2002; Ambrose et al. 2010; Kober 2015). Yet, a section about assessment is not included in this volume's chapters (Box 1.2). For better or worse, excluding assessment from the scope of each activity's description was intentional. Assessment (or grading) is a complicated endeavor with many subtle issues (e.g., see Driscoll and Wood 2007 and Dirks et al. 2014). Philosophies and practices vary widely among instructors and across course contexts. For each of the activities in this volume, student engagement, achievement of learning outcomes, and associated products (e.g., completed worksheets) could be evaluated and scored in numerous ways (if this is even desired). Because this book's purpose is to disseminate the activities which are, arguably, harder to devise, more space has been devoted to them at the expense of assessment and grading details which are left up to individual instructors to consider in context of their student populations, personal preferences, and syllabi. (As exceptions to this, some authors have suggested assessment (exam) questions or projects in the *Activities* or *Follow-Up Engagement* sections and provided example rubrics for scoring some student products in the supplementary materials.)

Box 1.2. Overview of Each Chapter's Structure and Content

1. *Introduction.* This section frames the larger context for the activity, explaining its connections to environmental, societal, and student-learning concerns. For brevity, it neither serves as a comprehensive overview of all related issues nor provides a thorough literature review. Authors cited only the most essential, relevant references, especially ones that can provide additional foundational information to readers with diverse disciplinary backgrounds. The general nature of the activity is briefly described so readers can start to envision its implementation. All chapters' introductions contain an explicit goal statement for the activity that speaks to the broader meaning, relevance, and implications of the intended student learning. Overall, each introduction frames the rationale and justification for why the reader should want to have their students complete the activity.

2. *Learning Outcomes.* Each chapter includes a set of three to five learning outcomes that the activity should help students achieve (see main text for context about outcomes). In true learner-centered form, outcomes are written with students as the subjects (not the course, goals, or content), followed by verbs that indicate what observable student actions would indicate learning gains (and can be assessed): "After completing this activity, students should be able to...." (This contrasts with the non-learner-centered emphasis in the phrasing "the goals of this activity are to....") Outcomes can be examined to quickly evaluate whether the activity might support an instructor's learning goals for a particular context. In addition, readers may discover that activities can be used to achieve other outcomes for other topics or goals not identified in the chapter.

3. *Course Context.* This section describes the context within which the author(s) developed and used the activity. Such information may be helpful for evaluating the activity's potential use, and possible needs for modifications, in the reader's teaching situation. For ease of reference, it is formatted as a bulleted list with the following information:

    (a) Basic characteristics of the course for which the activity was designed, including topic, educational level, majors or general education course, and class size
    (b) Expected duration in minutes (minimum or a range) and number of class meetings needed for the full activity
    (c) Background knowledge or other preparation needed by students, if any
    (d) A note about whether the activity could be adapted to (or has been used in) other course and learning contexts.

(continued)

Box 1.2. (continued)

4. *Instructor Preparation and Materials.* This section provides detailed, explicit, and precise guidance about what needs to be done before using the activity, as if it is the methods section of a scholarly research paper. In addition to basic preparation, most activities require that instructors review, and modify as needed for their own situation, resources that have been provided by the authors, such as student worksheets and example presentation slides. Preparation time varies among the activities and will in part depend on the instructor's preexisting knowledge about the activity's subject and desires to adjust the activity and provided materials.

5. *Activities.* In a step-by-step manner, this section describes precise instructions for both the instructor and students to complete the teaching and learning activities. From it, the reader should be able to clearly envision how the activity will proceed *before* implementing it (to facilitate preparation via "mental practice"). The detail and length of this section depend on the nature of the activity; as such, this section varies among the chapters more than other sections, with some using numbered or bullet-point lists for the steps. In all chapters, suggestions are provided for how instructors can guide students with background information, relevant summary talking points, clear instructions, and Socratic-style questions. This section is written for instructors; as relevant, some chapters provide separate instructions and worksheets in the online supplementary files that can be given to students.

6. *Follow-Up Engagement.* Suggestions in this section focus on how the activity's topic can be explored further in diverse ways, either by extending the activity directly (e.g., with discussion questions) or through complementary tasks (e.g., out-of-class assignments). In many cases, the suggestions could be adapted for formal assessment questions or projects (e.g., exam questions, essay prompts). To keep the chapters concise, the ideas in this section are presented as brief notes in bullet-list form and are not intended to be fully developed lesson plans or assignments; they serve to inspire readers to develop other teaching activities. When possible, other of the most relevant chapters in this volume are identified to suggest combinations of activities that can facilitate synergistic learning.

7. *Connections.* These notes indicate relationships among the topics of the chapter's activity and other environmental and sustainability study topics (especially those rooted in disparate disciplines). Sharing these connections with students can improve their abilities to synthesize information and create interdisciplinary mental models (which, as suggested by constructivist learning theory, can enhance their learning; e.g., Bransford et al. 2000;

(continued)

Box 1.2. (continued)

Ambrose et al. 2010). Other chapters in this volume that relate to the noted connections are highlighted to guide readers to additional relevant content and references.

8. *Electronic Supplementary Materials.* Contributing authors have been very generous in sharing resources associated with their teaching activities that enable their quick and easy adoption, including student worksheets, lists of additional background references, detailed instructor guides, example presentation slides, clickable links for online content (videos, articles), and answer keys. These materials are provided as electronic files on the chapters' websites which can be accessed from the book's main website. (Note that some materials needed for the activities are from copyrighted sources (e.g., articles, images) and could not be reproduced for cost or other reasons; their (online) locations or other information about how to access them should be indicated. However, if they cannot be found, please contact the author(s) or editor.)

9. *References.* For brevity, the reference list includes only core, essential references and does not reflect a comprehensive review of the chapter's topic(s). Especially for those less or unfamiliar with the topic(s), the suggested resources will provide foundational information to enable all instructors to gain the background knowledge needed to implement the activity. Many chapters list additional references in supplementary files to facilitate more in-depth exploration.

Peer Review of Teaching Activities

Higher-education professors are very familiar with the value and process of peer review for scholarly activity (e.g., for awarding grants and publishing articles). However, in general, instructors do not usually expect, or seek out, peer review of their teaching activities (unless these are formally submitted for publication, but it's probably fair to say that those who have done this are a small minority — one of the most common comments I received from authors was "I've never prepared a manuscript like this before"). Nonetheless, teaching activities can benefit substantially from critical peer review, as the editing process for this volume revealed. (Each chapter was first reviewed by two contributors of other chapters and the editor; subsequent reviews were made by only the editor.) The first drafts of the contributions varied widely in their levels of development and overall quality (in terms of content, the activity itself, and writing). Reviewers provided valuable comments about ways to improve each activity and its presentation in manuscript form; even the most straightforward ones and those with the highest-quality first drafts were strengthened by feedback and revisions. The most common improvements were made regarding the clear articulation of learning outcomes; increasing the specificity and explicit guidance in the descriptions for how to prepare for and implement the activity so that all readers could understand it easily; and including a variety of topics and perspectives in the *Follow-Up Engagement* and *Connections* sections. Reviewer feedback sometimes provided new ideas for alternative approaches, new connections, or simple modifications to the activity that could enhance student engagement or deepen learning gains. Inevitably, such revisions led to stronger, more rigorous and sophisticated activities that are likely to be more successful at fostering student learning.

One of the most critical aspects of the peer-review process (and one that I especially emphasized as editor) was providing feedback to authors about how to better align all parts of their manuscripts and activities to enhance their internal coherence and focus. Ensuring that lessons have well-aligned goals, activities, and concepts is a challenging yet crucial part of teaching (as is emphasized in the backward design approach described above; Wiggins and McTighe 2005). This is because, in part, an intentionally designed lesson with associated activities that clearly "make sense" allows students to easily see its logical organization and work in a more focused and enjoyable way toward achieving the outcomes (Bransford et al. 2000; Ambrose et al. 2010; Davis and Arend 2013; Kober 2015), rather than struggle as they try to figure out "what's the point?". Many chapters in this volume were significantly improved with revisions that brought all the sections and aspects of the activity into better focus and alignment (e.g., by removing points and components that were tangential and irrelevant to the focal goals and outcomes and created confusion about the focus). Hopefully, as a result, both the writing and activities are much stronger overall.

Editing this volume has convinced me that the processes of writing manuscripts about teaching activities and peer-reviewing them are both enormously valuable. All teachers should devote more time to doing both because both can help instructors become more thoughtful and critical about their own pedagogy—which ultimately makes them better teachers. With thoughtful review, all teaching activities can be improved—sometimes subtly, sometimes dramatically—to make them more coherent, rigorous, engaging, and, thus, effective. In turn, dissemination of highly polished, well-aligned, peer-reviewed teaching activities benefits the whole educational community by enhancing the pool of instructional resources that can be easily adopted. My personal perception is that the higher-education community in general—and the environmental and sustainability studies one especially—currently lacks sufficient opportunities, outlets, and infrastructure to facilitate these processes. Hopefully, as the predominant educational paradigm continues to shift toward one focused on learning, more instructors will devote more attention and time to helping each other peer-review, improve, and share our learner-centered teaching activities.

The Joy and Necessity of Learner-Centered Teaching for Environmental and Sustainability Studies

Although this chapter has highlighted some challenges of learner-centered teaching, I hope it convinces you (if you weren't already convinced) that, on balance, the approach benefits outweigh its costs, and the associated trade-offs are worthwhile to make. If any doubt or hesitancy remains, I hope that you will at least review and implement some of the wonderful activities in this book. They exemplify many of the benefits of learner-centered pedagogy, especially its enormous potential for fostering significant learning outcomes. (I used several of them while finalizing the book with positive results.) Personally, I remain committed to this approach despite its challenges because of my many encouraging experiences with it, including hearing appreciative student feedback. One key benefit that keeps me committed to it is the joy—the deep feeling of fulfillment that comes with meaningful accomplishment—associated with creative learner-centered teaching and seeing students engaged in active learning (a point also emphasized by Davis and Arend (2013)).

For teachers, joy can be enhanced by shifting from a teaching to a learning paradigm (Barr and Tagg 1995) because of how the latter frames and informs the work of teaching (as reviewed above). Instead of lecturers (and maybe discussion leaders), learner-centered instructors are designers of learning scenarios and inspirational and motivating guides who assist and challenge students to learn deeply while completing learning activities (Weston 2015). Taking on these roles opens the door for instructors to create more unique, varied, fun, thought-provoking, exhilarating, memorable, and—ultimately—rewarding and effective teaching activities. Further, as students are working in the classroom, instructors can facilitate their serendipitous A-Ha! learning moments through small-group and one-on-one discussions. Such interactions allow the observation of real-time student learning which yields smiles and genuine enjoyment from teaching. In my experiences, learner-centered pedagogies are much more likely than lectures to create such satisfying moments. Thus, using more of them can enhance the overall joy of teaching and generate more reminders about why educators love the teaching profession.

Similarly, learner-centered teaching activities can increase students' enjoyment of learning experiences. This is because, as teachers know well, the work and outcomes of successful learning can produce joy, exemplified by the pleasure of one's "mental light bulb" going off from a new insight. (Whether or not students are reflective and mature enough to recognize this type of joy in the moment is another question.) Since significant learning (sensu Fink 2003) is facilitated by learner-centered pedagogies, they are more likely to cultivate joy than non-learner-centered approaches. In part, this can be attributed to how they help create dynamic, energized classrooms that keep students awake, attentive, talking, writing, and thinking deeply and critically (rather than bored, drowsy, and distracted by their electronic devices, superficially listening and passively taking notes). Further, laughter and other positive social interactions can be wonderful results brought about by teaching activities, especially those arising from serendipitous and unique teachable moments. Although cultivating joy may not be a focal goal when designing teaching activities, their potential to do this suggests another valid reason to use them more often. Arguably, more joyful students are likely to be more interested, cooperative, and successful learners.

The abilities of learner-centered teaching to engage students point to their necessity for promoting excellent and joyful learning. Beyond this, the larger societal context of education should also guide pedagogical choices. In particular, the extraordinary environmental and sustainability challenges confronting humans around the world demand responses from educational systems (Orr 2004; MEA 2005). A first educational response is to increase efforts to help students more deeply construct transdisciplinary understanding of the diverse information (vocabulary, principles, questions, etc.) within environmental and sustainability studies (as reflected by the wealth of topics across this volume's chapters). Further, a critical educational goal must be to help students improve the skills and mindsets that will enable and motivate them to become socially and environmentally responsible and engaged problem-solvers and citizens. In particular, capacities for critical self-reflection are needed for them to recognize and honestly evaluate their own environmentally related beliefs, choices, and behaviors. Simply put, a teaching-centered paradigm is inadequate to achieve these crucial learning outcomes, especially given the high societal risks associated with failing to help students achieve higher levels of environmental literacy. As reviewed in this chapter, the best available scientific evidence indicates that learner-centered teaching approaches have higher efficacy for helping students achieve long-lasting, meaningful, and significant learning gains. In context of environmental and sustainability studies, a strong learner-centered teaching paradigm is a necessity for helping students become highly knowledgeable and skilled sustainability leaders.

In sum, the future sustainability and well-being of humanity and the biosphere depend on the environmental and sustainability education community's adoption of learner-centered pedagogies (Barr and Tagg 1995). The teaching activities in this volume aim to inspire and enable instructors from across disciplines and diverse teaching contexts to create more effective learner-centered classrooms that improve students' environmental literacy. I hope they bring joy to you and your students as they help transform our education systems in ways needed to support humanity's transition to a more sustainable future.

References

Ambrose SA, Bridges MW, DiPietro M, Lovett MC, Norman MK (2010) How learning works: 7 research-based principles for smart teaching. Wiley, San Francisco
Andrews SE, Frey SD (2015) Studio structure improves student performance in an undergraduate introductory soil science course. Nat Sci Educ 44:60–68

Barkley EF (2010) Student engagement techniques: a handbook for college faculty. Jossey-Bass, San Francisco

Barr RB, Tagg J (1995) From teaching to learning: a new paradigm for undergraduate education. Change 27:12–26

Bland M, Saunders G, Frisch JK (2007) In defense of the lecture. J Coll Sci Teac 37:10–13

Blumberg P (2009) Developing learner-centered teaching: a practical guide for faculty. Jossey-Bass, San Francisco

Bransford JD, Brown AL, Cocking RR (eds) (2000) How people learn: brain, mind, experience, and school. National Academy Press, Washington, DC

Brewer C, Smith D (2011) Vision and change in undergraduate biology education: a call to action. American Association for the Advancement of Science, Washington, DC

Brown PC, Roediger HL, McDaniel MA (2014) Make it stick: the science of successful learning. The Belknap Press, Cambridge, MA

Cardinale BJ, Emmett Duffy J, Gonzalez A, Hooper DU, Perrings C, Venail P, Narwani A et al (2012) Biodiversity loss and its impact on humanity. Nature 486:59–67

Carmi N, Arnon S, Orion N (2015) Transforming environmental knowledge into behavior: the mediating role of environmental emotions. J Environ Educ 46:183–201

Davis JR, Arend BD (2013) Facilitating seven ways of learning. Stylus, Sterling, VA

deWinstanley PA, Bjork RA (2002) Successful lecturing: presenting information in ways that engage effective processing. New Direct Teach Learn 89:19–31

Dirks C, Wenderoth MP, Withers M (2014) Assessment in the college classroom. Freeman and Company, New York

Doyle T (2011) Learner-centered teaching: putting the research on learning into practice. Stylus, Sterling, VA

Driscoll A, Wood A (2007) Developing outcomes-based assessment for learner-centered education. Stylus, Sterling, VA

Fink LD (2003) Creating significant learning experiences. Jossey-Bass, San Francisco

Freeman S, Eddy SL, McDonough M, Smith MK, Okoroafor N, Jordt H, Wenderoth MP (2014) Active learning increases student performance in science, engineering, and mathematics. PNAS 111:8410–8415

Gregory JL (2013) Lecture is not a dirty word: how to use active lecture to increase student engagement. Int J High Educ 2:116–122

Handelsmen J, Miller S, Pfund C (2007) Scientific teaching. W.H. Freeman and Company, New York

Hora MT (2014) Limitations in experimental design mean that the jury is still out on lecturing. PNAS 111, E3024. doi:10.1073/pnas.1410115111

Jensen JL, Kummer TA, Godoy PD (2015) Improvements from a flipped classroom may simply be the fruits of active learning. CBE Life Sci Educ 14:ar5. doi:10.1187/cbe.14-08-0129

Kober N (2015) Reaching students: what research says about effective instruction in undergraduate science and engineering. The National Academies Press, Washington, DC

MacKeachie WJ, Svinicki M (2006) MacKeachie's teaching tips. Houghton Mifflin, Boston

Malik AS, Malik RH (2012) Twelve tips for effective lecturing in a PBL curriculum. Med Teach 34:198–204

McBride BB, Brewer CA, Berkowitz AR, Borrie WT (2013) Environmental literacy, ecological literacy, ecoliteracy: what do we mean and how did we get here? Ecosphere 4:67, http://dx.doi.org/10.1890/ES13-00075.1

McNeil JR (2000) Something new under the sun: An environmental history of the twentieth-century world. W. W. Norton & Company, New York, NY

(MEA) Millennium Ecosystem Assessment (2005) Ecosystems and Human Well-being: Synthesis Report. Island Press, Washington, D.C.

Orr DW (2004) Earth in mind: on education, environment, and the human prospect, 2nd edn. Island Press, Washington, DC

Sapiains R, Beeton RJ, Walker IA (2015) The dissociative experience: mediating the tension between people's awareness of environmental problems and their inadequate behavioral responses. Ecopsych 7:38–47

Shephard K (2008) Higher education for sustainability: seeking affective learning outcomes. Int J Sust High Educ 9:87–98

Vincent S, Bunn S, Sloane L (2013) Interdisciplinary environmental and sustainability education on the Nation's Campuses 2012: curriculum design. The National Council for Science and the Environment, Washington, DC

Walker JD, Cotner SH, Baepler PM, Decker MD (2008) A delicate balance: integrating active learning into a large lecture course. CBE Life Sci Educ 7:361–367

Weimer M (2002) Learner-centered teaching: five key changes to practice. Jossey-Bass, San Francisco

Weston A (2015) From guide on the side to impresario with a scenario. Coll Teach 63:99–104

Wiek A, Withycombe L, Redman CL (2011) Key competencies in sustainability: a reference framework for academic program development. Sust Sci 6:203–218

Wiggins G, McTighe J (2005) Understanding by design, 2nd edn. Pearson, Upper Saddle River, NJ

Worthen (2015) Lecture me. Really. The New York Times. http://www.nytimes.com/2015/10/18/opinion/sunday/lecture-me-really.html. Accessed 19 Oct 2015

Part I
Conceptual Foundations and Frameworks

Chapter 2
Which Is Most Sustainable? Using Everyday Objects to Examine Trade-Offs Among the "Three Pillars" of Sustainability

Corina McKendry

Introduction

In everyday conversation, the term "sustainability" is used so frequently and in so many different contexts that, for some, it has lost precise meaning (Freyfogle 2006; Farley and Smith 2014). For others, the term signifies "environmental sustainability" and efforts to reduce the impact of human activities on the nonhuman world (see Caradonna 2014 for a useful discussion of the differences between sustainability and environmentalism). Though common, this view differs substantially from the word's meaning when it first gained prominence in the 1980s, as world leaders tried to find a workable balance between concerns about environmental degradation and the ability of people to meet their material needs (WCED 1987). Because of its conceptual fuzziness, everyone should carefully think about what they mean when they say something is "sustainable." In an introductory sustainability studies class, students should become familiar with the assertion that sustainability must incorporate the "three pillars" of social, economic, and environmental well-being. It is also important for them to understand the limitations of this conceptualization, particularly the challenges in reconciling tensions among the "three pillars" (Gibson 2006; Kuhlman and Farrington 2010; Morrison-Saunders and Pope 2013; also see Chaps. 23 and 24 in this volume).

Electronic supplementary materials: The online version of this chapter (doi:10.1007/978-3-319-28543-6_2) contains supplementary material, which is available to authorized users.

C. McKendry (✉)
Political Science Department and Environmental Program, Colorado College, Colorado Springs, CO, USA
e-mail: Corina.McKendry@ColoradoCollege.edu

© Springer International Publishing Switzerland 2016
L.B. Byrne (ed.), *Learner-Centered Teaching Activities for Environmental and Sustainability Studies*, DOI 10.1007/978-3-319-28543-6_2

The primary goal of the learning activity described below is for students to gain a basic understanding of sustainability that includes social, economic, and environmental aspects. The secondary goal is for them to see the difficulty of operationalizing this concept, especially when weighing potential trade-offs among these three pillars. To achieve these goals, students rank common household objects by what they perceive to be the most to the least sustainable and discuss their assumptions behind these rankings. Because the activity requires no student preparation, it works well at the beginning of a course or sustainability lesson. The activity's small-group discussions also help students talk to each other, thus building a sense of collaboration and community. Although this activity has elements of an informal life cycle assessment and could introduce a formal life cycle analysis project, its main purpose is simply to introduce students to the concept and challenges of sustainability.

Learning Outcomes

After completing this activity, students should be able to:

- Explain the "three-pillar" concept of sustainability.
- Question their previous assumptions about the meaning of sustainability.
- Articulate their own analysis of why an object is or is not sustainable.
- Describe the difficulty of weighing potential trade-offs among social, environmental, and economic aspects of an object's sustainability.

Course Context

- Originally designed as a lab activity for an introductory engineering and general education course and has been used in political science courses with 10–30 students
- 40 min (or longer with extended analyses and discussion) in one class meeting
- Requires no student preparation or background knowledge
- Adaptable to courses of any size

Instructor Preparation and Materials

To implement this activity, the instructor should be prepared, in the context of example objects and in a short summary lecture, to (A) explain the "three-pillar" conception of sustainability and (B) discuss the challenge of trade-offs when conflicts or tensions arise among the three pillars (e.g., the benefits of affordable necessities for low-income people may result in negative environmental consequences). Brief descriptions and illustrations of the "three-pillar" concept can be found on the US EPA website (u.d.) and in the introductory chapter of Caradonna (2014). Relevant perspectives are also provided by AtKisson (2013). See Kuhlman and Farrington (2010) for a useful critique of the "three-pillar" conception. Discussions of the challenges of trade-offs among the three pillars in the context of sustainable development can be found in Lele (1991) and Boström (2012)).

In the classroom, a writing surface such as a chalk or dry-erase board is needed to record student groups' sustainability rankings of the objects and the reasons for those rankings. Before class, the instructor should gather six to eight household objects that are likely to elicit a wide variety of initial, perhaps obvious and conflicting environmental, social, and/or economic associations. For example, a cell phone may be seen as not environmentally sustainable because of the materials used to manufacture it and the growing problem of electronic waste. However, cell phones can be very valuable for improving banking and health care in developing countries, therefore contributing significantly to social and economic sustainability. Other objects and potential questions and trade-offs for each are suggested in Electronic Supplementary Material (ESM) A. Obtaining detailed information for each item is not necessary to successfully complete this activity. However, instructors who want to know as much as possible about their chosen objects can find information on many common household products from the Worldwatch Institute (2004), www.goodguide.com, and www.madehow.com. For more detailed assessments of a smaller number of objects, see Pearce (2009) and Leonard (2011).

Activities

To begin, the instructor shows the objects to the students. Without any other prompting or discussion, the instructor asks them to individually rank the objects from the most to the least sustainable and write down their rankings. (This takes about 5 min.) Starting the activity without opening remarks is effective because an important aspect of the exercise is for students to realize that others may have very different ideas as to what objects are more or less sustainable. However, a brief introduction can be added based on the instructor's preference and course context. As they are determining their rankings, students can be invited to come up to examine the objects more closely. After everyone completes their ranking, the instructor places the students into groups of three or four and instructs them to share their rankings and discuss the reasons for their choices. Each group should then use a process of debate and discussion to create one collective ranking of the objects. (This should take about 10 min.)

Next, the instructor asks a group to volunteer their ranked list of objects, writes this on the board, and then asks the group to explain the rationale for their decisions. These reasons should be recorded on a different section of the board. Depending on how long the instructor wishes to spend on the activity, the group can explain the rank of each object or just those ranked most and least sustainable. Additional groups should be invited to explain their rankings and rationales, in turn, while the instructor records them on the board. This can be done for all groups or for as many as time permits or are needed to provide sufficient information for the subsequent discussion. As more rankings and rationales are recorded, students will probably find that other groups had very different rankings from their own, accompanied by compelling reasons for them. This is likely to cause group explanations and class discussion to shift toward more nuanced analyses about what makes something sustainable. As potential debates emerge, the instructor can allow the discussion to proceed organically (allowing students to respond to points made by those in other groups) or guide it by pointing out notable talking points, as when one object is ranked as highly sustainable by one group and low by another. The instructor may also encourage groups to elaborate on their explanations by asking the class if anyone agrees or disagrees with a particular criterion offered by a group. After the first few groups have listed their sustainability rankings, the important element of the activity is not the rankings but the discussion of the criteria. If new criteria for sustainability emerge from the discussion, the instructor should add these to the board.

In the author's experience, the most common initial criterion students use to rank objects is their (presumed) environmental sustainability. They usually focus on whether the object is recyclable or reusable; some students will raise questions about the environmental impact of the production process. Even with this narrow view of sustainability, the divergent rankings quickly lead students to see that there are different ways of thinking about whether or not something is environmentally sustainable (e.g., is the spray bottle sustainable because it is reusable and mostly recyclable or is it not sustainable because it is made from plastic?). Depending on the class and students' knowledge, some groups might raise points that address economic and social sustainability. If these aspects do not emerge from the discussion, the instructor should ask questions to raise these issues (see ESM-A for examples). Either way, varied aspects of social and economic sustainability (and environmental as needed) should be emphasized to help challenge and deepen students' initial views. After discussing several examples of how the three aspects of sustainability (or lack thereof) and possible trade-offs among them can be seen in the objects, the instructor should present a short lecture on the "three-pillar" conception of sustainability and its importance, providing summary points and referencing the example objects and notes on the board as appropriate (see the references cited above for additional content and perspectives).

To conclude, the instructor can encourage students to offer additional examples of ways in which the objects may meet one or two of the pillars of sustainability but not another (e.g., natural play dough from the local, small business may be environmentally and economically sustainable, but it is also very expensive. Who can afford this play dough? Does its expense make it less socially sustainable?). There is no need to come to definitive answers about these potential trade-offs. The goal is simply to help students recognize that, as important as the "three-pillar" understanding of sustainability is, it can be hard in practice to balance and find synthesis among the economic, social, and environmental aspects of sustainability.

Follow-Up Engagement

- These questions can be used to further examine the concept of sustainability:
  - What does it mean for something to be "sustainable?"
  - What information about an object would you need to be more confident in your assessment of its sustainability? Do you think that information would be hard to get? Why or why not?
  - What are the relationships among environmental, social, and economic sustainability? How do these objects show potential tensions among these three pillars? What should be done if the three pillars of sustainability are in conflict with each other? How can they be balanced? What should be prioritized and why? (Also see Chaps. 23 and 24.)

- Ask students to think of a brand-name food, personal care, or household item. Using the criteria and concepts that emerged from the activity, have them spend 10 min writing about the sustainability of that item. Their analysis should include what they already know about the item (e.g., it is certified organic) and what they do not know (e.g., the working conditions in the manufacturing plant). Then, ask students to look up the item at www.goodguide.com and write a two-page reflection on the provided information. The reflection can discuss whether their guesses about the item in their initial writing match the information from the website, the ways that the item is or is not sustainable, any trade-offs among the three pillars they see in the item, and how these trade-offs could be balanced to make the item more sustainable. (See Chap. 22, this volume, for a related activity.)
- The issues and questions raised in this activity can be revisited throughout a course. Asking students to reflect on the exercise is particularly useful when other examples of trade-offs between the pillars emerge from course material or when students revert to using the concept of sustainability to mean environmentalism. This exercise can also be revisited at the end of a course to encourage students to think about how their understanding of sustainability has changed.

Connections

- This activity provides concrete illustrations that can be referred to when discussing many environmental and sustainability topics, including economic growth and development, environmental policy, energy use, human health, consumerism

and advertising, and corporate social responsibility. To this end, instructors may want to keep a permanent record of the board notes so they can be revisited later.
- For courses that include a more in-depth and extended life cycle analysis or sustainability assessment, this activity establishes a useful foundation. For example, it can be used as a starting point for individual and group research projects in which students find data and reports pertaining to sustainability assessments of the objects used in this exercise (see the *International Journal of Lifecycle Assessment*, Klöpffer 2003, and Chaps. 3, 22, and 23 of this volume).
- This exercise can be referred to when systems thinking is discussed in a course in that it requires an examination of multiple multifaceted processes and the complexity of their interactions.

Acknowledgments This activity was developed in association with the Sustainability Engineering & Ecological Design group at the University of California, Santa Cruz and with funding from the US National Science Foundation (CCLI/TUES Awards # 0837151; 0187589; 1023054).

References

AtKisson A (2013) Sustainability is for everyone. ISIS Academy Publishing, Oxford
Boström B (2012) A missing pillar? Challenges in theorizing and practicing social sustainability: introduction to the special issue. Sustain Sci Pract Pol 8:3–14
Caradonna JL (2014) Sustainability: a history. Oxford University Press, Oxford
Farley HM, Smith ZA (2014) Sustainability: if it's everything, is it nothing? Routledge, New York
Freyfogle ET (2006) The sustainability sham. Orion 25:11
Gibson RB (2006) Beyond the pillars: sustainability assessment as a framework for effective integration of social, economic and ecological considerations in significant decision-making. J Environ Assess Pol Manag 8:259–280
Klöpffer W (2003) Life-cycle based methods for sustainable product development. Int J Life Cycle Assess 8:157–159
Kuhlman T, Farrington J (2010) What is sustainability? Sustainability 2:3436–3448
Lele SM (1991) Sustainable development: a critical review. World Dev 19:607–621
Leonard A (2011) The story of stuff: the impact of overconsumption on the planet, our communities, and our health – and how we can make it better. Free Press, New York
Morrison-Saunders A, Pope J (2013) Conceptualizing and managing trade-offs in sustainability assessment. Environ Impact Assess Rev 38:54–63
Pearce F (2009) Confessions of an eco-sinner: tracking down the sources of my stuff. Beacon, Boston
U.S. EPA. (u.d.) Sustainability primer. http://www.epa.gov/ncer/rfa/forms/sustainability_primer_v7.pdf. Accessed 12 Mar 2015
WCED (1987) Our common future. Oxford University Press, Oxford
Worldwatch Institute (2004) Good stuff? http://www.worldwatch.org/system/files/GoodStuff Guide_0.pdf. Accessed 31 Mar 2015

Chapter 3
An Introduction to Systems Thinking Using Plastic Dinosaurs

Laura A. Hyatt

Introduction

All the manufactured objects that surround us are the products of a complex set of interacting systems (Meadows 2008; Pearce 2008). The raw materials that are used to make these objects must be extracted, transported, and processed in energy-intensive ways. People who design these objects and work with machines to make them have resource needs derived from cultural systems that may be different from our own. Design elements are selected on the basis of economic considerations, aesthetic choices, and mechanical constraints. After use, many objects are disposed of in ways that may or may not recapture some of the materials and energy they embody. These diverse, interconnected systems need to be examined if we are to redesign them to maximize the triple bottom lines of environment, equity, and economics (Braungart and McDonough 2002; Leonard 2010; also see Chaps. 2, 23 and 24 of this volume).

The activity described below engages students in collaboratively constructing a concept map depicting the connections among many of the natural and human systems involved and resources used in the production of a common toy, a plastic dinosaur. The choice of a plastic dinosaur is intentional, for it explicitly connects our manufacturing systems with nature's carbon cycling system. The polymers in plastics are derived from compounds in petroleum and natural gas (American Chemistry Council 2005), and the carbon in these raw materials was fixed by photosynthesis about 300 million years ago (before dinosaurs existed, roughly 200 million years ago). That carbon is now being returned to the atmosphere through industrial combustion which is causing climate change. Thus, a toy representing an extinct animal

Electronic supplementary materials: The online version of this chapter (doi:10.1007/978-3-319-28543-6_3) contains supplementary material, which is available to authorized users.

L.A. Hyatt (✉)
Department of Biology and Sustainability Studies Program, Rider University,
Lawrenceville, NJ, USA
e-mail: lhyatt@rider.edu

37

L.B. Byrne (ed.), *Learner-Centered Teaching Activities for Environmental and Sustainability Studies*, DOI 10.1007/978-3-319-28543-6_3

that is made using raw materials and energy derived from extinct plants and animals provides a focal point for introducing students to systems thinking.

As an introductory activity, this informal life-cycle assessment (Curran 1996) prompts students to begin to regard everyday objects as products of systems, all of which have impacts on the environment, economics, and human well-being. By encouraging students to deeply consider the interconnections that contribute to one manufactured object, this activity provides students with the tools to analyze and evaluate sustainability challenges from a more holistic perspective. Further, it fosters collective learning because it draws on every student's experience, knowledge, and understanding. It frames a system-focused course in sustainability very nicely because it has tremendous staying power; the dinosaur made of dinosaurs provides a concrete symbol that can be referred to again and again throughout the semester to foster deep and memorable learning.

Learning Outcomes

After completing this activity, students should be able to:

- Articulate a long list of interacting, overlapping systems and their parts that are involved in the manufacture of everyday objects.
- Trace the interconnecting paths taken by energy, people, and materials involved in manufacturing processes of common consumer products and their environmental relationships.
- Approach new problems using critical systems thinking, especially with respect to interacting systems.

Course Context

- Originally used for an introductory, college-level Introduction to Sustainability Studies course with 10–20 students
- 20–120 min in one class meeting
- No background preparation is needed on the part of students
- It has been used in small groups for a municipal sustainability organization, at mixed-age elementary school Earth Day events and with large groups of high-school students for a day-long symposium on environmental sustainability

Instructor Preparation and Materials

Obtain a set of plastic toy dinosaurs, preferably with packaging, although one plastic dinosaur will work. The instructor should become familiar with processes involved in plastic manufacture (ACC 2005), how carbon is fixed by photosynthesis and released through cellular respiration (i.e., from a general biology textbook), and the creation and extraction of fossil fuels (USDOE 2015).

A large (at least 2 m^2) writing surface and markers are needed to help students visualize the interconnections among the systems and materials that emerge from the conversation. A large chalk or whiteboard works well, although very large paper sheets and markers are useful to preserve the diagram long term. Good background reading on systems thinking (and associated key terms) for the instructor is provided in Meadows (2008). Leonard (2010) and Braungart and McDonough (2002) provide insights as to how our manufacturing and supply chain principles can be troublesome and articulate a vision of how they can be changed. These three references could be used as pre-activity or follow-up reading assignments to spark additional student learning and discussion. Additional background resources about life-cycle analyses and systems thinking are provided in Electronic Supplementary Materials (ESM) A.

Activities

The instructor opens the activity by displaying the toy dinosaur(s) and asking "What went into getting this plastic dinosaur into our classroom today?" A dinosaur can be drawn on the board as the starting point for an informal concept map. To encourage (or require) participation by all students, the instructor can invite everyone to write down five answers to the question and then call on people to share their thoughts. As students make contributions, the instructor adds them to the concept map, connecting resources to form interconnected systems. These maps can become very large and complex (see ESM-B for an example). If desired, instructors can invite students to use smartphones or computers to look up answers to questions that arise as the map is developed. The instructor's role in this activity is as a guide, asking questions, urging participation, and scribing the ideas of the group.

It can be expected that, depending on their backgrounds, experiences, and perspectives, different groups of students will focus on different aspects of plastic-dinosaur production, leading to varied maps. In the author's experiences, while some have focused on travel and transport, others were interested in the idea of a dinosaur as a toy. Other groups worried about materials and packaging, while others focused on embodied energy or on the workers whose lives intersected with the toy. Suggested avenues to encourage students to explore are provided in Box 3.1. A key goal for the analysis is to include people, material resources, environmental resources, and the flow of money in the students' considerations of the dinosaur.

At the conclusion of the students' work, the instructor can help identify themes and wrap up the activity with guiding questions and reviewing key issues as needed. It is also helpful to look for patterns and themes, steering participants toward excitement about the connections and systems involved, and to avoid becoming overwhelmed with their sheer numbers. Toward the end of the class, the instructor can backtrack through some of the pathways the materials and energy have taken, highlighting common connections in the systems involved; part of the point of sustainability is that all of our systems are interconnected. Highlighting those connections during this activity can help students develop the habit of thinking about them throughout a course (and beyond).

Box 3.1. Questions to Consider About Plastic Dinosaurs and Suggested Answers

1. Moving objects:

   (a) Where are many plastic toy factories located? *Asia.*
   (b) How did it get here from there? *Distribution centers, shipping vessels, tractor trailer trucks, toy stores, and cars.*
   (c) What are they made of? *Plastic, metal, rubber, and glass.*
   (d) Where do they get their energy? *Electricity—coal/nuclear/natural gas/oil.*
   (e) What kind of packaging is it sold in and how is that made? *Cardboard and paper from trees, shrink-wrap, and plastic from natural gas-based synthesis process;* see ACC (2005).

2. Making objects:

   (a) What is plastic made of? *Polymers made from petroleum and natural gas; see ACC reference.*
   (b) How did this toy get its color or shape? *Molds for pouring melted plastic made elsewhere, inorganic pigments, and dyes with metal ingredients.*
   (c) Where does petroleum/natural gas come from? *Middle East, Geology.*
   (d) Where does the carbon in petroleum come from? *Photosynthesis/carbon fixation.*
   (e) Where did all the atoms inside this plastic dinosaur come from? *Big bang.*
   (f) What are factory machines made of? *Cast metal.*
   (g) Where does that material come from? *Mines, which require more machinery to extract.*
   (h) Where does the energy to run the machines come from? *Coal, hydroelectric, and nuclear power.*
   (i) How does it get to the factory? *Consider electricity distribution grid infrastructure.*
   (j) What will happen to this dinosaur when we are tired of it? *Landfill.*
   (k) What might happen to the dinosaur once it reaches a landfill? *Might break into smaller pieces but no biodegradation will happen.*
   (l) What about the packaging that the dinosaur came in and all the "extras" generated in manufacture (gas emissions, plastic filings, extra dyes, packaging parts, etc.)?

3. People and objects:

   (a) Who sold this item? *Consider minimum wage, labor laws, and vendors.*
   (b) Who manufactured the item? *Factory workers, their needs.*

(continued)

> **Box 3.1.** (continued)
>
> (c) Who manufactured the tools to make the item? *More factories and manufacturing.*
> (d) Who packaged the item?
> (e) What resources do they consume? *Food, water, electricity, heat, and transportation.*
> (f) Who paid the people who made this item?
> (g) Who designed this toy? *Paleontologists? Artists?*
>
> 4. Why this object?
>
> (a) What scientific discoveries occurred (when) to make us think of making and using a dinosaur replica as a toy? *Concept of extinction developed by Cuvier in 1800s related to discovery of mastodons, considered heretical. Dinosaurs per se were not understood to be extinct reptiles as we know them now until the middle of the nineteenth century.*
> (b) What is the message when a parent sends to a child when presenting this item as a toy (about nature, imagination, science)? *Consider value of natural history and natural spaces in our lives and transmission of values from parent to child and child-to-child.*
> (c) Why is this made from plastic and not some other material? *Durability, changes in technology,* etc.

This activity could be used with larger classes (>20 students) in two ways. The class could be split into groups with each developing their own diagram, reserving time at the end of class to share and combine them. Alternatively, each group could complete the diagram but analyze the dinosaur with a different question such as what role did people play in making this dinosaur? Where and how was money exchanged? Where was energy needed? What natural resources were consumed? There will inevitably be overlapping among the groups, which can enhance the take-home message that "nothing is an island" and that each manufactured object embodies small bits of energy and materials from many overlapping and nested systems.

Follow-Up Engagement

- Ask students to reflect on what they learned from the exercise, either orally or in writing.
- Have students independently create their own life-cycle assessment diagram for another common object, either assigned by the instructor or selected themselves, adding specific researched and quantitative details dependent on the focus of the course.

- Have teams of students invent a toy that could perform the functions of a plastic dinosaur, but be sustainable in its production, use, and disposal.
- If the students generate a set of themes, use them throughout the term as a template to examine other topics and engage in other problem-solving activities.

Connections

- This activity is a simple version of a life-cycle analysis. It is helpful to remind students of this opening activity if this concept is addressed later in the term. For instance, cradle-to-cradle design principles involve reimagining the entire production and disposal process. By analyzing existing production and disposal processes, students should see a bigger picture of how they may be modified. Labor practices and fair trade economics are sometimes addressed in corporate responsibility statements. This exploration of production makes these invisible issues more visible. (See Chap. 22, this volume, for a related exercise.)
- Because it involves plastic and dinosaurs, this activity can introduce students to the carbon cycle, fossil fuels, and climate change and how these are connected. The environmental impacts of extractive technologies for fossil fuels can be integrated into these discussions as well.
- Human factors play an important role in why we do things the way we do. Consumerism, satisfaction, and happiness are just as, if not more important, in driving our economy and how we elect to allocate our resources. (See Chap. 22, this volume, for related topics.)

Acknowledgments This activity was developed with the inspiration of Ralph Copleman, Executive Director of Sustainable Lawrence from 2006 to 2011.

References

American Chemistry Council (ACC), Plastics Industry Producer Statistics Group (2005) How plastics are made. http://plastics.americanchemistry.com/Education-Resources/Plastics-101/How-Plastics-Are-Made.html. Accessed 1 Jul 2015

Braungart M, McDonough W (2002) Cradle to cradle: remaking the way we make things. North Point Press, New York

Curran M (1996) Environmental life-cycle assessment. McGraw-Hill, New York

Leonard A (2010) The story of stuff: the impact of overconsumption on the planet, our communities and our health – and how we can make it better. Simon & Schuster, New York

Meadows D (2008) Thinking in systems: a primer. Chelsea Green Publishing, White River, VT

Pearce F (2008) Confessions of an EcoSinner: tracking down the sources of my stuff. Beacon, Boston

United States Department of Energy (USDOE) (2015) How fossil fuels were formed. http://www.fossil.energy.gov/education/energylessons/coal/gen_howformed.html. Accessed 1 Jul 2015

Chapter 4
An Introductory Examination of Worldviews and Why They Matter for Environmental and Sustainability Studies

Loren B. Byrne

Introduction

A worldview—the set of beliefs, values, and attitudes inherent in every person—is a complex cognitive and affective lens through which a person sees and interprets the world, including human-environment relationships. Although worldview lenses may often be latent and unconscious, they exert strong (though not exclusive) influence on people's decisions and behaviors (Newell et al. 2014). Examining worldviews—what they are, how they're formed, and their implications—is relevant to environmental and sustainability studies, in part, because divergent worldviews create conflict about environmental problems and solutions (Hedlund-de Witt 2012). For example, whether or not a person accepts scientific evidence about anthropogenic climate change—and supports actions to do something about it—is often determined more by political beliefs than the evidence itself (Joyce 2010, Kahan et al. 2011; also see Chap. 39 of this volume). Understanding the myriad dimensions of worldviews and their diversity within and between cultures (e.g., Hedlund-de Witt et al. 2014, Saucier et al. 2015) is important as a means to break down stereotypes, better understand the roots of disagreements, and (hopefully) more successfully cultivate the respect, compassion, unity, and constructive civil dialogue needed to find common ground, compromise, and make positive changes for a more sustainable society and environmental outcomes (e.g., Chaps. 19 and 29 of this volume).

To this end, the activity described below introduces students to worldviews and their links to environmental and sustainability issues through informal, open-ended,

Electronic supplementary materials: The online version of this chapter (doi:10.1007/978-3-319-28543-6_4) contains supplementary material, which is available to authorized users.

L.B. Byrne (✉)
Department of Biology, Marine Biology & Environmental Science, Roger Williams University, Bristol, RI, USA
e-mail: lbyrne@rwu.edu

© Springer International Publishing Switzerland 2016
L.B. Byrne (ed.), *Learner-Centered Teaching Activities for Environmental and Sustainability Studies*, DOI 10.1007/978-3-319-28543-6_4

exploratory dialogue, catalyzed by the examination of four individuals' hypothetical worldviews as playful case studies. The activity's goal is to help students realize that everyone has a unique worldview that emerges from personal experiences, such that we should resist temptations to make assumptions about them or judge them too quickly. As a simple introductory activity, it can be adapted to align with instructors' backgrounds and course goals and can be expanded with instructor's in-class guidance to include discussions about related issues (e.g., stereotypes, prejudice, empathy, cultural diversity, public debates) that may arise when analyzing personal concerns such as beliefs, values, and experiences in a social-environmental context.

Learning Outcomes

After completing this activity, students should be able to:

- Describe what a worldview is and its many dimensions.
- Ask questions that help reveal dimensions of a person's worldview.
- Discuss how worldviews influence people's environmental and sustainability perspectives.
- Explain why examining worldviews is a valuable part of environmental and sustainability studies.
- Reflect on one's own worldview and the factors that have shaped it.

Course Context

- Developed for an introductory-level general sustainability studies class with 15–30 students from diverse majors and all class levels
- 20–25+ min in one class meeting
- No student preparation or background is needed; a brief instructor-led introduction can be provided, and if desired, students could be asked to do a preparatory reading
- Adaptable for use in classes of any size and can be extended to 50+ min by including more in-depth content, analysis, and discussion

Instructor Preparation and Materials

The instructor should be comfortable with introducing the concept of worldviews (including providing a definition such as those in Electronic Supplementary Materials (ESM-A)) and guiding students in a brainstorming exercise to (1) list their major thematic components, (2) discuss factors that influence them, and (3) explain how they relate to environmental and sustainability issues. As presented below, the activity asks students to contribute information about these three issues, so the instructor should be open to integrating many different student responses into the discussion. (For this purpose, a chalkboard or whiteboard, or other writing surface, is needed in the classroom to record student contributions.) However, it may be helpful to prepare an outline of key points that the instructor wishes to emphasize and can use to fill in the discussion as needed. Some basic information is suggested in the Activities section below and in ESM-A. Recommended background articles include Koltko-Rivera (2004) and Hedlund-de Witt (2012) who provide introductory overviews, Saucier et al. (2015) who report results and belief statements from a global-scale survey of worldviews that provide helpful context, and Newell et al. (2014) who discuss worldviews in relation to environmentally related decision-making and behaviors. For those who wish to adapt the activity with more content, formality (e.g., using more organized frameworks), or specific information (e.g., published data or case studies), the articles cited above and additional ones listed in ESM-A can be consulted for more information. In addition, a set of presentation slides that can be used (or adapted as desired) to introduce worldviews (including a simple conceptual framework on slide 8) and guide discussion during and following the activity is provided in ESM-B (expanding discussion using all the slides can fill a 50-min session).

The materials needed are (at least) four pairs of "funny eyeglasses" for students to wear; in the activity, these provide a physical representation of the metaphor that a worldview is a "lens" that a person looks through. Such inexpensive glasses are often in the toy and party sections of dollar and other discount stores, but finding used ones or borrowing them would align better with sustainability values. Possible types to obtain include giant "clown" sunglasses, those with eyes that droop down on springs, any that have funny or scary eyes as lenses, and "nerd" glasses with thick lenses (see a photo of examples in ESM-C). In the author's experience, such funny glasses engage students because they are unexpected, engender laughter, and create a lighter classroom atmosphere (perhaps balancing seriousness of the subject); however, inexpensive standard reading glasses could be used to make the same points.

Alternatively to using purchased, manufactured glasses, the instructor could ask students to create their own "funny" eyeglasses using recycled or reused materials and unique designs. This would make the activity more learner centered, add additional dimensions to the discussion (e.g., we each construct our worldviews using information from our environments), and allow all students to wear glasses during the discussion. (It also reinforces sustainability values by avoiding the use of consumer "stuff"; however, using mass-produced glasses could catalyze discussions about this issue to enrich the lesson.) This option may work well in large classes in which the activity is adapted for use in small discussion groups, creating a need for more glasses.

Finally, the instructor is encouraged to review the activity description below and reflect on how it might be tailored to her specific teaching situation, preferences, and student population. In particular, the framing of creating hypothetical identities might be approached differently with more diverse populations and advanced classes (e.g., in environmental psychology, sociology, or anthropology) than at an introductory level focused on general sustainability in a less diverse class. Because worldviews are complex and individualistic and relate to examining people's personal, sometimes private, characteristics and identities, discussing them in classes, especially in an informal, open-ended way as suggested below, may carry some risk of bringing forth sensitive topics and perspectives. As such, preparing for the lesson should include considerations of how to (1) avoid reinforcing stereotypes, (2) not allow any prejudiced statements to go unquestioned, and (3) overall, ensure all students will feel comfortable participating in the conversation. However, in the author's experiences using this as an introductory activity over six years, no concerns or unwanted complications have arisen; rather, it has consistently yielded engaging, thoughtful, and productive conversations that informed further conversations throughout a course.

Activities

The instructor should introduce the topic of worldviews as desired depending on the nature of the course and its goals, the placement of this activity within a larger lesson, and students' educational level(s). (The author has provided a very brief introduction and definition (see slides 3–5 in ESM-B) and then used the activity to catalyze a student-centered discussion to examine details.) The instructor then asks for four volunteers (or more as desired) to line up in front of the whiteboard/chalkboard. (It is up to the instructor whether or not to say why the volunteers are needed; the author chooses not to so it is a surprise.) These students will be given hypothetical "identities" and worldviews using the following steps.

The instructor asks the rest of the class to suggest possible categories (or themes) of worldview components and/or factors that influence a worldview. (In the author's experience, listing components, influences, or both will yield a similar discussion and learning outcomes. It may be difficult to disentangle them and is not necessary for purposes of this brief activity but can be done to more formally structure the discussion if desired.) Common responses may include religious beliefs, ethnicity, hometown, family size, education level, and career, but students should be encouraged to brainstorm as many as possible including less familiar and obvious ones (e.g., views about citizenship rights and responsibilities). The instructor should then guide students to choose (collectively or via individual requests) four or five (or more) component/influence categories from among the student-generated ideas (and instructor-contributed ones as needed) that will be written in a vertical list on the board.

Next, this list is used to create a hypothetical (and partial) "identity" for each volunteer by asking the class to provide a specific characteristic for each of the worldview categories for each volunteer, while ensuring a diversity of traits is provided among the identities for each category. Each characteristic is written on the board to create an "identity" list for each volunteer. (See Table 4.1 for examples and

Table 4.1 Example of list of worldview components/influences and hypothesized identities for two student volunteers

Worldview component/influence	Volunteer 1 "identity"	Volunteer 2 "identity"
Age	26	55
Religion	Christian	Buddhist
Hometown	New York City	Nepalese village
Environmental ethic	Ecocentric	Anthropocentric

ESM-C for a photo of an in-class result.) This could be done category by category or person by person. If the latter, students should be encouraged to avoid creating stereotyped identities in favor of creating unique, somewhat unpredictable combinations of traits for each individual. Key talking points that can be brought forth after this step include (1) recognizing that worldview components and factors influencing them are numerous and diverse and that (2) because each person has a unique background (i.e., set of experiences and context for his or her life), it can be argued that everyone develops and possesses a unique worldview, aspects of which may not always follow from preconceived stereotypes (e.g., the notion of Republicans who "hate" environmental causes can be challenged by examples of those who have pro-environmental stances).

For these two brainstorming steps, allowing open-endedness and randomness is acceptable and can lead to serendipitous talking points, unexpected and unusual combinations of "identity" characteristics (that may even seem implausible), and beneficial teaching moments and insights, alongside cultivating a learner-centered classroom. However, as the instructor feels necessary and comfortable, the activity can proceed with stronger instructor-led guidance toward preselected choices that result in more intentional results (e.g., to highlight or avoid certain issues).

Once the volunteers' "identities" are established, the instructor hands each volunteer one of the funny eyeglasses to put on (or asks them to put on the ones they made) and explains that these lenses represent the holistic worldview that arises from each individuals' set of characteristics listed on the board (see photo in ESM-C). To foster class discussion, the instructor can ask the class to (1) summarize and synthesize the overall nature of each volunteer's "identity," (2) compare and contrast them, and (3) hypothesize how the components/influences might (hypothetically) individually and collectively affect the lens (glasses) that the person is looking through and (4) how that could impact their perspectives on environmental and sustainability issues. (In ESM-B, slides 10–14 pertain to relationships between worldviews and environmental issues.) Alternative interpretations should be asked for to emphasize that there may be multiple possibilities and that we should resist temptations to make simplistic assumptions about a person's views based on known or perceived characteristics. To this end, it may be prudent to suggest to students that it's ok to answer questions with "I don't (or can't) know" and honestly admit that "it's impossible to know what a person would think because we haven't lived her life." Related to this, these discussion questions could be posed: what additional information would you like to have about each of these people? How would you ask them about their views in civil, nonjudgmental, and nonthreatening ways? In an iterative way, throughout the dialogue, the instructor could ask the simple questions "Why? In what ways? What makes you say that?" to bring forth more critical reflection, discussion, and analysis that help deepen students' insights.

In this activity, the glasses provide a concrete visualization of the abstract worldview concept. To utilize them further, the instructor can ask the volunteers to take off their glasses, switch them, and comment on how the "world" looks different without any glasses and then through the lenses of someone else's worldview (i.e., based on the "identities" on the board that accompany each set of glasses; again, this is not to suggest that we can "know" or should assume what it is like to look through someone else's views; in part, the point is to recognize that we really can't do this easily if at all). This "changing of one's lenses" can help make the point about the value of, first, recognizing that we have a worldview that colors our own outlooks and, second, trying to step outside our own worldviews to try to understand others' points of view (i.e., developing empathy, walking in others' shoes). A summative question to follow from this might be, why can it be helpful to identify it characteristics of a person's worldview when discussing environmental and sustainability issues with them? Further talking points to consider include the more that we recognize how our own and others' worldviews frame and limit our perspectives and hinder openness to new ideas, the more likely we might be to understand, navigate, and overcome the fundamental root causes of our differing perspectives—and perhaps even find common ground. Such outcomes cannot be guaranteed of course; deconstructing worldviews might also risk inflaming tensions and disrespect. However, the more we can see each person's worldview and identity it as a complex whole with a diverse range of interacting ideas that emerges from a unique set of experiences, the more we might be able to understand a person's views and productively engage her in civil dialogue about environmental and sustainability problems and solutions.

Hopefully the activities and discussion above cause students to begin to reflect on their own worldviews. Instructors could facilitate this throughout the discussion by asking them to share reflections about their own beliefs, values, and attitudes and what has influenced them. As desired and time permitting, a valuable way to ensure this outcome is for students to write a brief statement about their own worldviews (i.e., a minute paper) and then share reflections with the whole class. They could be asked to focus on the list of categories on the board or describe how a specific topic in the course (e.g., previewing a topic for a future class) appears through their worldview.

Follow-Up Engagement

• Additional discussion can be fostered with these questions:

   – In sum, why are there so many diverse beliefs and worldviews in society? Given the diversity of worldviews, how can we engage in constructive dialogue to promote sustainability? (These are included on slides 15 and 16 in ESM-B.)

- What are the dominant influences on and beliefs within a worldview? How will these vary across people, space, and time?
- Is it possible for two people to share (nearly) the same worldview?
- Can a person's worldview change over time? Why might it?
- How and why might some people's worldviews cause them to not accept environmental problems as true or serious?
- Can someone's worldview (or parts of it) be wrong? Are all worldviews valid?

- Have students examine quantitative analyses (e.g., Saucier et al. 2015) and narrative, first-person descriptions (e.g., TIB n.d.) that reveal people's diverse worldviews, possibly focusing on dimensions that were identified in the class activity.
- Assign students a more in-depth, formal writing assignment in which they explain major components of and influences on their own worldviews (e.g., TIB n.d., Jurin and Hutchinson 2005).

Connections

- Throughout a course, ask students to analyze a topic from diverse worldviews (i.e., by changing the lenses they're looking through; see Chaps. 29 and 39 of this volume).
- When introducing scientific thinking and/or principles, compare and contrast scientifically informed worldviews with those based on religious or other belief and value systems.
- Ask students to read diverse texts about a controversial environmental topic that reveal divergent opinions and speculate on the components of the authors' worldviews that influenced their assumptions and conclusions about the issue (e.g., Chaps. 36, 37, and 41 of this volume).
- The issue of worldviews can be discussed in context of disciplinary topics from environmental psychology (e.g., why our behaviors may not align with aspects of our beliefs and values), economics (e.g., an unquestioned belief that economic growth is good), and communications (e.g., strategies for effective interpersonal and intercultural dialogue).

References

Hedlund-De Witt A (2012) Exploring worldviews and their relationship to sustainable lifestyles: towards a new conceptual and methodological approach. Ecol Econ 84:74–83

Hedlund-De Witt A, de Boer J, Boersema JJ (2014) Exploring inner and outer worlds: a quantitative study of worldviews, environmental attitudes and sustainable lifestyles. J Environ Pschol 37:40–54

50   L.B. Byrne

Joyce C (2010) Belief in climate change hinges on worldview. NPR's all things considered. http://
    www.npr.org/templates/story/story.php?storyId=124008307. Accessed 12 Mar 2014
Jurin RR, Hutchinson S (2005) Worldviews in transition: using ecological autobiographies to
    explore students' worldviews. Environ Educ Res 11:485–501
Kahan DM, Jenkins-Smith H, Braman D (2011) Cultural cognition of scientific consensus. J Risk
    Res 14:147–174
Koltko-Rivera ME (2004) The psychology of worldviews. Rev Gen Psychol 8:3–58
Newell BR, McDonald RI, Brewer M, Hayes BK (2014) The psychology of environmental deci-
    sions. Annu Rev Environ Resour 39:443–467
Saucier G, Kenner J, Iurino K et al (2015) Cross-cultural differences in a global "Survey of World
    Views.". J Cross Cult Psychol 46:53–70
This I Believe (TIB). http://thisibelieve.org/. Accessed 27 Jul 2015

Chapter 5
Building Resilience: Modeling Resilience Concepts Using Legos

Courtney E. Quinn and John E. Quinn

Introduction

A simple definition of resilience is the ability or capacity of an object, individual, or system to recover from pressure, change, a challenge, or disturbance. Studies of resilience focus on individuals, social systems (e.g., cities), ecosystems (e.g., lakes), and coupled social–natural systems (Duckworth et al. 2007, Liu et al. 2007, Scheffer et al. 1993, Stone-Jovicich 2015). For example, diverse farming systems are suggested to be more resilient than monocrops because they are less sensitive to fluctuations in climate and economic markets (Lin 2011). Resilience provides a model to understand changes to human and natural systems and for redesigning current, and creating new, institutions and ecosystems. Importantly, resilience is applied across disciplines and thus serves as a unifying concept in the study of human–environment relationships and their sustainability.

The activity described below engages students in thinking about resiliency by constructing models using toy building blocks (e.g., Legos). It was inspired by a passage in Andrew Zolli's book *Resilience: Why Things Bounce Back* (2012), in which he discusses characteristics of resilient systems (e.g., modularity and redundancy) and how to bolster resiliency of systems by allowing systems to reorganize in the face of stress and perturbations. He suggests that "…reorganization [of systems] is made feasible by certain structural features of resilient systems. While these systems may appear outwardly complex, they often have simple internal

Electronic supplementary materials: The online version of this chapter (doi:10.1007/978-3-319-28543-6_5) contains supplementary material, which is available to authorized users.

C.E. Quinn (✉)
Sustainability Science, Furman University, Greenville, SC, USA
e-mail: courtney.quinn@furman.edu

J.E. Quinn
Department of Biology, Furman University, Greenville, SC, USA
e-mail: john.quinn@furman.edu

© Springer International Publishing Switzerland 2016
L.B. Byrne (ed.), *Learner-Centered Teaching Activities for Environmental and Sustainability Studies*, DOI 10.1007/978-3-319-28543-6_5

51

modular structure with components that plug into one another, much like Lego blocks, and- just as important- can unplug from each other when necessary" (p. 11).

The goals for this learning activity are for students to construct and explain a physical model (with Legos or similar building blocks) that demonstrates their understanding of the traits and characteristics of resilient systems. By doing so, they should gain a greater conceptual understanding of resilience while also creating concrete mental models of resilience concepts to reference in future work.

Learning Outcomes

After completing this activity, students should be able to:

- Define "resilience" and provide examples related to individuals and social–ecological systems.
- Explain key concepts of resilience.
- Apply resilience concepts to build a model using toy building blocks that they can explain and defend.
- Compare and contrast building-block resilience models with individual, social, and ecological examples.

Course Context

- Developed for an undergraduate sustainability science course for 15–20 students
- 30–45 min in one class meeting
- Before the activity, students read provided material (depending upon course topic) and/or should be provided some background about resilience
- Adaptable to courses with 5–30 students and any course level that includes a discussion of resilience. Can be modified for longer or shorter durations depending upon student ability, engagement in activity, or available class time

Instructor Preparation and Materials

To complete this activity, the instructor should be able to explain and discuss concepts of resilience and facilitate students' creative thinking on resilience concepts (see Box 5.1 for brief overview of concepts and Electronic Supplementary Materials (ESM-A) for background and readings). The instructor may want to prepare a lecture to precede the activity and assign readings based upon the focus and level of the class. For example, Holling (1973) is considered a classic introductory paper; it may not be suitable for introductory classes but should be reviewed by the instructor for background and would be suitable for upper-level or graduate classes. Fazey et al. (2007)

Box 5.1. Characteristics of Lego Models That Reflect Resilience in Systems

Characteristic or related concept	Model applications
Grit	The ability to continue on in an activity despite obstacles (Electronic Supplementary Materials (ESM-B, Photo 1)
Redundancy	Having three pillars hold up a second level when only two are needed
	Easy to remove one Lego and replace it with another without structure falling over
	To increase redundancy building walls, have Legos overlap rather than stand in a single tower (ESM-B, Photo 2)
Alternative stable states/thresholds/hysteresis	Thresholds: How many 2×2 bricks can you stack before it falls over?
	Alternative stable states: Can any blocks or parts be rearranged while the overall structure or function of the model remains the same and stable?
	Relate the fact that you cannot just take the last brick off and have the tower stand again to the idea of hysteresis (ESM-B, Photo 3)
Recover from crises/disturbance	Can their structure rock/sway and still stand?
Uncertainty	How many bricks are needed to have redundancy?
	How might the threshold for your tower vary as a function of other factors?
	What would happen if you removed a brick?

review resilience in social–ecological systems. Folke et al. (2010) review the history of resilience and present multiple applications in social–ecological systems. In addition, the Stockholm Resilience Center (n.d.) has a clear summary and video at their website.

This activity works best as a culminating activity to apply and connect key concepts and test students' understanding of them. The materials needed are a large bucket of toy building blocks (e.g., Legos or similar generic ones); the more and diverse the pieces, the better. The authors have successfully used a two-gallon bucket full of Legos for 15 students. Building blocks can be obtained cheaply at garage sales, thrift stores, etc., or even borrowed from a neighbor. Examples of models are shown and described in ESM-B.

Activities

The activity starts with the instructor standing with a large tub of building blocks at the front of the room. The blocks may be left out during class lecture or discussion, but to build anticipation, do not tell students what they are for. To frame the activity, end the lecture with Zolli's quote, from the introduction above, written on the board or projected in a presentation slide.

To introduce the activity, share the assignment guidelines and requirements on a projected slide or written on the board (see ESM-C for example slides). Specifically, ask students to create a model (in 10–20 min) that demonstrates a characteristic of resilience through its structure (Box 5.1). (It is recommended that Box 5.1 is not shown at first, but it could be used as a concluding slide.) Ask the students to work alone or in pairs to ensure all students are highly involved in both conceptualization of the idea and building the model. Emphasize that there is no "right" way to build a model. If students feel stuck or are uncertain about how to proceed, picture of example models could be shown to help inspire them (e.g., ESM-B).

The instructor should move around the room while students build and ask them questions regarding their conceptualization of resilience. For example, what concept related to resilience did you decide to model and why? Why did you include this base/piece/link? Could you include another piece to make your structure more resilient? Ask them to try and push their structure over; is it resistant to toppling or does it fall and stay together or break? If students do not have much or any background in resilience thinking, they may need more reassurance about their work or more help identifying the concepts they want to illustrate. The instructor can challenge students' thinking with more in-depth questions. For example, ask them to demonstrate a second resilient characteristic with additional pieces or integrate the concept of uncertainty by modeling how a threshold (for toppling over) might change if various factors of their model are changed such as height, base, width, etc.

After the building activity is completed, discuss with students what makes building blocks like Legos particularly good for creating resilient structures. Discuss aspects of resilience that such blocks demonstrate and refer again to the Zolli (2012) quote above. Then, students can share their work with the class. Ask students which characteristics of resilience they incorporated. If a sense of competition is of value for the class, a vote can be taken for the best model or the model that best withstands a disturbance (i.e., knock the model over or remove a random brick). To summarize the activity, review the characteristics students applied to their models and any characteristics they did not incorporate. It is a good idea to take a picture of each model for future use, either on exams or for future classes.

Follow-Up Engagement

- Ask students how they would improve upon their own or another's design to make it more resilient.
- Discuss why building with Legos creates a more resilient structure than building with other materials such as regular, non-interlocking children's blocks.
- Ask them to make connections across models by discussing interconnectedness across coupled human and natural systems and how this relates to local and global resilience in the "real world."

- Discuss whether individuals or social–ecological systems need to have a certain number of characteristics to be resilient.
- Have students research and connect their models to examples of resilience in nature (e.g., Yellowstone after fires), social systems (e.g., New Orleans after Hurricane Katrina), or individuals (e.g., Nelson Mandela). Ask them to share and discuss their findings about characteristics of resilience or lack of resilience.

Connections

This activity provides context for many subsequent topics that might arise in a course such as the following:

- In the context of current events, a large natural or social disaster could be used as an opportunity to apply concepts explored in students' models to discussion of the social–ecological system's resilience or lack thereof.
- As resilience is the ability or capacity of an object, individual, or system to recover from pressure, change, a challenge, or disturbance, it has many topical connections that might arise in a general sustainability or environmental studies course. For instance, resilience could be discussed in context of ecology lessons, political events, or even an individual's life challenges like a death in the family or disease. Social topics such as food, water, energy, population growth, access to education and jobs, gender issues, and political representation can all be related to resilience of human groups. Ecological topics such as forest fires, habitat loss, pollution, biodiversity loss, and climate change will all touch upon resilience of natural and human systems. (For related topics, see Chaps. 12, 13, and 31 of this volume.)

References

Duckworth AL, Peterson C, Matthews MD et al (2007) Grit: perseverance and passion for long-term goals. J Pers Soc Psychol 92:1087–1101

Fazey I, Fazey JA, Fischer J et al (2007) Adaptive capacity and learning to learn as leverage for social–ecological resilience. Front Ecol Environ 5:375–380

Folke C, Carpenter SR, Walker B et al (2010) Resilience thinking: integrating resilience, adaptability and transformability. Ecol Soc 15:20

Holling CS (1973) Resilience and stability of ecological systems. Annu Rev Ecol Syst 4:1–23

Lin BB (2011) Resilience in agriculture through crop diversification: adaptive management for environmental change. BioScience 61:183–193

Liu J, Dietz T, Carpenter SR et al (2007) Complexity of coupled human and natural systems. Science 317:1513–1516

Scheffer M, Hosper SH, Meijer ML, et al. (1993) Alternative equilibria in shallow lakes. Trends Ecol Evol 8:275–79

Stockholm Resilience Center (n.d.) The best explanation of resilience. http://www.stockholmresil-
    ience.org/21/research/what-is-resilience.html. Accessed 9 July 2015
Stone-Jovicich S (2015) Probing the interfaces between the social sciences and social-ecological
    resilience: insights from integrative and hybrid perspectives in the social sciences. Ecol Soc
    20:25
Zolli A (2012) Resilience: why things bounce back. Simon & Schuster, New York, NY

Chapter 6
Eco-crimes and Eco-redemptions: Discussing the Challenges and Opportunities of Personal Sustainability

Mercedes C. Quesada-Embid

Introduction

Environmental and sustainability studies can create pathways for positive social transformation through fostering individual reflection and behavioral change. The circumstances in which many people find themselves, however, are not always conducive to immediate change in support of sustainability. Gifford (2011) and Gifford and Nilsson (2014) explore a wide range of variables, such as childhood experience, cultural background, and personality, that may affect our ability, knowledge, and/or willingness to behave in pro-environmental ways (also see Steg and Vlek 2009 for discussion of actual and perceived barriers to behavioral change). There is much navigating and negotiating that takes place in personal decision-making, which can lead to unintended contradictions between people's pro-sustainability values and their unsustainable actions.

The term "green guilt" refers to the emotional sentiment associated with an awareness of not making choices according to what is best for the environment and sustainability when one wants to. Bedford et al. (2011) suggest that oftentimes, guilt by itself is not a fruitful mechanism for fostering sustainable behaviors, unless opportunities for compensatory action are available. Thus, reflecting on one's unsustainable behaviors should be coupled with a focus on one's abilities to engage in pro-sustainability actions. Although some may question whether individual actions can make a worthwhile difference (e.g., Maniates 2001), Steg and Vlek (2009) argue that individuals can contribute significantly to long-term societal

Electronic supplementary materials: The online version of this chapter (doi:10.1007/978-3-319-28543-6_6) contains supplementary material, which is available to authorized users.

M.C. Quesada-Embid (✉)
Sustainability Studies Department, Colorado Mountain College, Edwards, CO, USA
e-mail: membid@coloradomtn.edu

sustainability by adopting pro-environmental behaviors. How to encourage individuals to consistently reflect upon and change their actions remains challenging, but it is clear that individuals have myriad opportunities to take meaningful and impactful steps toward reducing their ecological footprints (e.g., see Dietz et al. 2009 and Gardner and Stern 2008).

In this chapter's activity, students identify personal behaviors that could be considered "crimes," against sustainability (that would generate green guilt) and contrast them with "eco-redeeming" acts which cultivate sustainability. When coupled with class discussion, this activity encourages students to assess personal behaviors through a lens of sustainability; fosters critical thought and reflection; acknowledges conscious and unconscious participation in particular behaviors; provides an opportunity for students to evaluate their own actions, as well as those of society at large; and contemplates the implications of both for positive and negative sustainability outcomes. Although it can induce some discomfort because it requires critical self-reflection, in the author's experiences, students have always had a very positive response to this activity, in part, because it bridges the personal with the academic. Overall, the goal is to engage students in open and creative dialogue about how to promote positive social change through attention to individual behaviors. For a variety of courses, this insightful, exploratory activity can help students become excited about the possible solutions and behaviors that should/could be commonplace in society, if we all operated with more environmentally-ethical and sustainable awareness.

Learning Outcomes

After completing this activity, students should be able to:

- Evaluate personal and societal behaviors in relationship to positive ("eco-redemptions") and negative ("eco-crimes") sustainability outcomes.
- Assess the prominence of green guilt and other personal and societal difficulties of sustainability.
- Articulate the need for creative, positive behaviors and communicate possible "eco-redemptions" to exemplify that change.
- Understand the value of nonjudgmental reflection, critical inquiry, and thoughtful dialogue as constructive ways to inspire an individual's capacity to improve sustainability practices.

Course Context

- Often used for a foundations of sustainability course consisting of environmental/sustainability majors. However, it was originally designed for an advanced class on the ethics of sustainable development, comprised of business and economics students not majoring in environmental/sustainability studies

- 50 min but can be extended with longer discussion and more students
- No preparation by the students is expected. This activity works well as an ice-breaker, primer, or introductory activity when students are not familiar with one another or their collective thoughts on sustainability. (This allows for the anonymity of the contributions to be upheld with more ease.)
- Adaptable for any class size with instructor-facilitated, whole group discussions

Instructor Preparation and Materials

The instructor will need to acquire enough small, blank scraps of paper for each student to have two (e.g., $3 \times 5''$) and one pencil or pen for each student (if pens are used, they should all be the same color to support anonymity). The instructor will also provide two labeled bowls (or other containers) in which to place the written "eco-crimes" and "eco-redemptions" that students share. The instructor should prepare to facilitate a discussion surrounding the importance of examining individual behavioral choices in the context of sustainability and environmental issues. This should include defining key concepts and terms and guiding a dialogue on "eco-crimes" and "eco-redemptions" (see context and references in the introduction and additional resources in Electronic Supplementary Materials (ESM-A)). "Eco-crimes" generally encompass actions that we do that are less sustainable and "eco-redemptions" are those actions that are more sustainable. (See additional discussion points and questions in the activities section below.) The instructor's ability to draw associations between the students' behaviors and the broader implications of those behavioral choices will help to enhance the activity's capacity to encourage reflection. For example, students may find the Greendex survey (from National Geographic; see the website links in Electronic Supplementary Materials-B) to be an intriguing set of findings on the emotions of green guilt that accompany over-consumptive practice and how/why there are distinctions among the ecological footprint of different nations.

Activities

The instructor begins by introducing the context of the activity to students and delineating the theoretical parameters of how individual choices can impact sustainability in both positive and negative ways. For instance, our individual behaviors impact aspects of our environment at varying levels, and if we want to become more sustainable as a society, we need to begin to honestly examine them. However, it is not always possible for our ideals to match our reality and oftentimes trade-offs are necessary. All members of society participate in a range of different behaviors including some that we would not want to showcase, or simply do not want others to know about, and others that we are very proud of when it comes to environmental and sustainability awareness. These comprise our "eco-crimes" and "eco-redemptions."

Box 6.1. Examples of "Eco-crimes" and "Eco-redemptions"

Eco-crimes

- Littering
- Eating at a fast-food restaurant
- Not buying organic or fair trade, when able
- Driving instead of biking/ walking/carpooling
- Using toxic household cleaners
- Buying bottled water instead of a filter for tap
- Purchasing new instead of second hand
- Using an electric clothes dryer in warm months
- Mistreating/harming others and other species
- Taking long, hot showers

Eco-redemptions

- Boycotting an unethical company
- Volunteering at a senior center
- Utilizing cloth napkins
- Planting native trees/shrubs/ grasses
- Starting an organic community garden
- Defending another's rights
- Lowering the thermostat and layering up
- Buying cruelty-free personal products
- Getting books from the public library
- Using a reusable water bottle

Instructors should introduce summative, simple definitions, such as *eco-crimes are things we do that hinder sustainability and "eco-redemptions" are things we do that help increase sustainability*. The "eco-crimes" would be acts that they wish had not occurred and/or that they would like to see end. "Eco-redemptions" are behaviors that we would like to see more of. The use of "eco" in this activity is in the whole sense of the term, meaning the "home"; it is inclusive so as to expand to all aspects of sustainability and not limit the activity only to conceptions of environment/ecology/ natural resources. (See Box 6.1 for examples that can be shared with students to aid their understanding and inspire their thinking; however, their contributions should be creative and distinct from the examples provided by the instructor.) The students should understand that these behaviors, no matter how frequently performed (once or many times as habits), have an impact on sustainability and that the overall goal of this exercise is to assess the opportunities for our personal behaviors to be more sustainability-enriching.

Students should be told that they will be tasked with writing down their own "eco-crimes" and "eco-redemptions." Students should allow the "eco-crime" discussion to inspire and inform their approach to the "eco-redemption" portion, but please remind students that the "crimes" and redemptive acts they share should not correspond to one another directly (i.e., contributions to each bowl should not be paired so as to cancel each other out). For example, "Eco-Crime: Failed to recycle; Eco-Redemption: Recycle more" is not what the activity calls for. The crimes and redemptions should be distinct from one another and imaginative.

Moreover, students should be told that the instructor will be calling on everyone during the course of the activity so as to ensure as equal participation as possible, especially if it is being used as an icebreaker early in the course.

Sustainability can be challenging at times to talk through with students because it is a very personal field; thus it is important to cultivate spaces where individuals can be honest about the contradictions and hypocrisies that sometimes create disjuncture between our ethics and behaviors. It is important for instructors to reiterate to students the tenets of good classroom and academic demeanor. Students should be reminded of the importance of kindness, respectfulness, and patience when they are participating in the activity. There should be no judgment passed on, or belittling remarks made about, the "eco-crimes" that students choose to share. The classroom should be a safe space where trust and camaraderie abound. It is helpful for instructors to state that all have participated knowingly and unknowingly in behaviors that have negatively impacted other people and species, locally and otherwise.

Instructors should emphasize the significance of anonymity by reminding students that it is necessary for everyone's comfort so that students can provide honest and specific responses without concern for being judged. (Related to this, students can be given the option of sharing acts that they witnessed someone else commit. This small caveat can make some students feel more comfortable, as it adds another layer to the ability to remain anonymous.) During the activity, it is recommended that instructors keep an informal but scholarly tone, sharing personal vignettes (or well-known cultural references) as appropriate to help students navigate the dialogue and understand its directionality and trajectory. Making positive remarks about each of the "eco-redeeming" acts shared by students during the discussion and limiting any directly negative remarks about the "eco-crimes" shared are also good approaches. Finally, instructors should prepare themselves for the layered unpredictability of the contributions, as this activity is very open ended and can go in a variety of directions. These recommendations apply to both of the discussions described below.

Students will begin by writing their "eco-crimes" on a scrap of paper provided by the instructor, which will be folded and placed into the respectively labeled container that the instructor has prepared. The instructor shuffles the scraps of paper, and one by one, students will randomly remove one and read it aloud to the class. Discussion ensues regarding the "crimes," including why such behaviors are undesirable from a sustainability perspective, why individuals and society permit their continuance, and how best to avoid them in the future. A small discussion could be allowed one by one for each "crime," or it may be more conducive to allow them all to be shared before having a more global discussion. A benefit to commenting on each one is that all of the student contributions are honored equally and the "crimes," themselves are comparably assessed, regardless of their seeming (in)significance. In addition, during the discussions of each behavior, the class could be polled (anonymously, e.g., with clickers or scraps of paper or by quick show of hands) to obtain a sense of how many students have engaged in each behavior. Questions that may be useful for this discussion are:

- Do you think that individuals experience a kind of "green guilt" when they choose to participate in "eco-criminal" behaviors or is complacency too embedded in mainstream society for such guilt to exist?
- At times our behaviors are a result of a particular circumstance, in which we have little to no choice. In what instances do we sometimes need to negotiate or justify unsustainable behaviors despite our overall good intentions?

After the "eco-crime" dialogue, students will share an "eco-redemption." Just as with the crimes, these are written and placed in the corresponding container to then be shuffled and read aloud by the students. Discussion of these "eco-redeeming" acts should include an evaluation as to how they embody sustainability and promote mindful behaviors, as well as how they may provide opportunities by which individuals can help mitigate the ecological, economic, and social disparities that surround us. Questions that may be useful for this discussion are:

- What will it take for the dominant behavioral norms within society to become more of the "eco-redeeming" sort, rather than the "eco-criminal" sort? Explain your perspective.
- Why does it seem more difficult to engage in these redeeming behaviors more often? Are there any negative trade-offs that arise from these positive behaviors?

Once both sets of written contributions have been discussed, the dialogue should evolve to assess, address, and recapitulate the activity and the issues it raised as a whole. The discussion element is an important part of the activity as it is designed to provide a foundation for a reflective dialogue on the ease with which "eco-crimes" can be committed and the individual choices and societal behaviors that permit them. The discussion should seek out practical and effective means by which to foster more sustainable behaviors and ways to encourage responsible actors in society. Essentially, the focus of the discussion is a balance between reflecting on why "eco-crimes" happen and understanding why "eco-redemptions" are more advantageous to society and how the latter can be encouraged and facilitated. Although the content of each dialogue will depend on the specific contributions that students make, here are three questions that can help students begin the reflective process:

- For which set of behaviors—"eco-crimes" or "eco-redemptions"—was it easier to come up with an example? Elaborate on your response.
- As you consider the idea of personal sustainability, what are some of the most common unsustainable human behaviors that we need to unlearn at the societal level in order for sustainability to be successful?
- If you were to establish a set of criteria to determine the level of severity and level of goodness of the "crimes" and "redemptions," how would you go about doing so and what would it consist of?
- How can we avoid the quick categorizations of being too idealistic or too pessimistic when we discuss change toward sustainability while still remaining practical in scope?

- In what ways can an acknowledgment of green guilt help to stimulate thought about how best to align our personal and professional selves more fully with our desired value systems?

Follow-Up Engagement

- These questions could be used to extend the discussion:
  - What areas of study are involved with helping to foster sustainable behaviors? Why is it relevant and valuable to use interdisciplinary thinking and methods to study and change sustainability behaviors?
  - What relationships, if any, exist between eco-leadership and becoming an exemplar of "eco-redeeming" acts in society? (See Chap. 7 of this volume for more on leadership.)
- Students could keep a scholarly journal that explores personal observations of self and society, analyzing the interrelations of these topics and describing any behavioral trends that students notice, especially highlighting people and organizations that exhibit empowering examples of social strength, leadership qualities, and innovative pro-sustainability thought. This could be done for their local communities and regional and national governments, and at the international scale.
- Throughout the course, a box could be maintained into which students contribute ongoing ideas for "eco-redeeming" acts. This encourages positivity and allows for moments during the semester when redeeming behaviors can be read aloud to remind the class about the impact of positivity within environmental and sustainability studies.
- A version of this activity can be exercised at a campus Earth Day celebration or other sustainability event. Students can have a table set up with a box serving as the receptacle for attendees to come by and anonymously contribute an "eco-crime" while having another box with varied "eco-redeeming" acts prepared beforehand by students. Participants pull out a redeeming act with the hope that they will adopt this new behavior while reducing their engagement in the "eco-crime."

Connections

- Depending on the crimes and redemptions contributed by students, discussions could connect to many environmental and sustainability topics, such as systems thinking, climate change, food issues, overconsumption, ethics, environmental justice, fair trade, sense of place, politics, and effective communication (see the related chapters in this volume for activities about these topics).

- Further discussions about ecopsychology could enhance the theoretical understandings that underpin elements of green guilt, societal complacency, nature detachment, and barriers to behavioral change.
- Behavioral economics could provide a platform by which to explore the human tendency toward responsible and alternative consumption patterns and the philosophical ethics of egoism and altruism (see Chap. 22 of this volume).
- Civic engagement and the prevalence of social capital could give students a foundation by which to recognize the abundance of "eco-redeeming" acts taking place around them, fostering an ethic of reciprocity and community building.

References

Bedford T, Collingwood P, Darnton A, et al (2011) Guilt: an effective motivator for pro-environmental behaviour change? RESOLVE: Working Paper. p 1–24

Dietz T, Gardner GT, Gilligan J et al (2009) Household actions can provide a behavioral wedge to rapidly reduce US carbon emissions. Proc Natl Acad Sci 106:18452–18456

Gardner GT, Stern PC (2008) The short list: the most effective actions US households can take to curb climate change. Environ Sci Policy Sust Dev 50:12–25

Gifford R, Nilsson A (2014) Personal and social factors that influence pro-environmental concern and behaviour: a review. Int J Psychol 49:141–157

Gifford R (2011) The dragons of inaction: psychological barriers that limit climate change mitigation and adaptation. Am Psychol 66:290–302

Maniates MF (2001) Individualization: plant a tree, buy a bike, save the world? Global Env Pol 1:31–52

Steg L, Vlek C (2009) Encouraging pro-environmental behaviour: an integrative review and research agenda. J Environ Psychol 29:309–317

Chapter 7
Engaging with Complexity: Exploring the Terrain of Leadership for Sustainability

Matthew Kolan

Introduction

Given accelerated changes and increasingly complex challenges facing life on Earth, knowledge and expertise may not be the limiting factors in our collective ability to live well and create conditions for life to thrive. Instead, many scholars and practitioners (e.g.,Scharmer 2009, Senge 2008) have suggested that a sustainable future will require leadership skills, mindsets, and dispositions that are well suited for engaging with *complex systems*, defined in part by their dynamic, nonlinear, and unpredictable nature (Rickles et al. 2007; also see Chaps. 3, 10, 28, and 31 of this volume). This contrasts with traditional leadership practices, which often emphasize control, management, expertise, formulaic solutions, and universal "best" practices (Wheatley 2010). These traditional approaches can be effective when engaging with systems that are largely stable, predictable, linear, and distinct. However, they often fall short when used to address complex sustainability challenges that are dynamic and span disciplines, fields, and social–ecological dimensions.

Today's challenges call for differentiated approaches to leadership that recognize the limits of our knowledge, create the possibility for collaboration, and are flexible enough to respond to ever-changing conditions. As Snowden and Boone (2007, p. 75) suggest, complex challenges necessitate leadership practices that "can help generate ideas, open up discussion, set barriers, stimulate attractors, encourage dissent and diversity, manage starting conditions and monitor for emergence."

Electronic supplementary materials: The online version of this chapter (doi:10.1007/978-3-319-28543-6_7) contains supplementary material, which is available to authorized users.

M. Kolan (✉)
Rubenstein School of Environment and Natural Resources, University of Vermont, Burlington, VT, USA
e-mail: mkolan@uvm.edu

© Springer International Publishing Switzerland 2016
L.B. Byrne (ed.), *Learner-Centered Teaching Activities for Environmental and Sustainability Studies*, DOI 10.1007/978-3-319-28543-6_7

The group activity described below is designed to create a learning experience that aligns with Snowden and Boone's (2007) suggestions. It draws on the collective experience of the students, avoiding singular answers or predetermined solutions. The activity is intended for courses that explore or are connected to the diverse and varied terrain of *leadership for sustainability*. The activity is based on an essential question: what skills, mindsets, and dispositions might help leaders address increasingly complex challenges to support a healthier and more sustainable future? The goal of the learning activity is for students to explore this fundamental question and generate areas of inquiry that can be further explored in future course content and activities.

Learning Outcomes

After completing this activity, students should be able to:

- Identify unique characteristics of complex systems and complex social–ecological challenges.
- Examine mindsets and behavioral patterns that may limit their ability to effectively address challenges in complex systems.
- Identify areas of inquiry related to leadership for sustainability that merit further exploration.
- Reflect on how to apply their newfound understanding about complex systems to improve their own leadership skills for engaging with complex sustainability challenges.

Course Context

- Developed for an upper-level environmental problem-solving capstone course with 75–100 students
- 60–75 min in one class setting (depending on use of optional background materials)
- Background readings (suggested below) can be helpful but are not essential
- Easily adaptable to smaller (minimum 20 students) or larger groups (up to 300 students) and courses that address topics of leadership, sustainability, problem-solving, or systems thinking

Instructor Preparation and Materials

To facilitate this activity, the instructor should be prepared to define complex systems and identify unique characteristics of complex challenges using examples (Box 7.1; also see Meadows (2008), Rickles et al. (2007), and Ladyman et al.

Box 7.1. Four Types of Challenges and Associated Effective Leadership Practices

Simple challenges: the domain of known knowns • Repeating patterns are evident • Consistent events and stability • Clear cause and effect relationships • Single correct answers exist *Examples*: baking a cake, changing the oil in a car *Possible leadership practices*: sense, categorize, and respond; use best practices	**Complicated challenges**: the domain of known unknowns • Patterns are discernable • Cause and effect relationships are discoverable • More than one good solution possible *Examples*: building a skyscraper, designing a computer program *Possible leadership practices*: sense, analyze, and respond; solicit expert diagnosis; synthesize conflicting advice
Complex challenges: the domain of unknown unknowns • Nonlinear (small changes can create large effects) • Unpredictable • Lack of central control agent • Emergence of new properties *Examples*: responding to climate change, ecosystem management *Possible leadership practices*: probe, sense, and respond; create conditions for new possibilities to emerge; invite tension and creativity	**Chaotic challenges**: the domain of the unknowable • High turbulence • High tension • No clear cause and effect relationships *Examples*: responding to a highly contagious new disease outbreak or terrorist attack *Possible leadership practices*: act, sense, and respond; attend to immediate impacts; look for what is working

Adapted from Snowden and Boone (2007)

(2013) for more background). Preparing students with background readings (e.g., previous citations) can be helpful but is not essential. However, an introduction to the nature of complex systems (and how they differ from simple, complicated, and chaotic systems) should be offered to introduce the activity at the start of the class session. The instructor should also point out that these different conditions necessitate unique leadership practices. See Snowden and Boone (2007) (especially their table in "Decisions in Multiple Contexts: A Leader's Guide") for an excellent overview of this concept. Examples of the four types of challenges (provided in ESM-A for projection) can be expanded into one-page case studies to give to students (see section "Activities"). Additional resources about leadership for sustainability (provided in ESM-C) can be consulted for more background as desired. It can also be helpful for the instructor to have some familiarity with World Café group process methodology (Brown 2005, The World Café Community Foundation 2015) since that process is adapted as a central element of the group activity.

The central activity of this lesson is best done in a room with movable seats and tables to enable student mobility. Students will self-organize into groups of five and each group will need one large sheet of paper (e.g., $3' \times 4'$ or flip chart) and two markers. The instructor will also need index cards (one per student) and tape to post the large sheets of paper to the wall.

Activities

Depending on the students' prior experiences, the instructor can begin by introducing or reviewing characteristics of complex systems and complex challenges (~10 min). If time is short, this introduction can be offered as a mini-lecture that includes examples to illustrate the variations between different kinds of challenges and contexts (see Box 7.1). However, if more time is available, place students into small groups (of three or four) and assign each group one of the four "types of challenges" from Box 7.1. Give each group an example that illustrates their "type of challenge" (e.g., from Snowden and Boone 2007). Ask the students to generate a preliminary list of characteristics for their assigned challenge type. Review their responses with the whole class, and add to each list to ensure that everyone understands the distinctions among them (~15 min).

After the introduction, offer additional examples of sustainability challenges (e.g., structural racism, water pollution), and ask which type of challenge best describes each example and why. Explain that because most sustainability challenges take place within complex contexts, the following activity focuses on leadership practices needed to effectively engage with this complexity (~5 min).

This is a good time to introduce a simple conceptual framework for leadership. At a basic level, leadership is "the act of making an intervention within a system, to change the results that a system produces" (Smith 2015). It is important to acknowledge that complex challenges are not easily addressed via leadership approaches that rely solely upon expertise or best practices. (In the author's experience, many students believe that they are not prepared to address complex challenges unless they "know it all.") Rather, the unpredictability of complex challenges may require a different set of leadership skills to navigate ambiguity and uncertainty (~5 min).

Next, the instructor introduces the lesson's essential goal and question (see section "Introduction")and acknowledges that the insights generated through the activity will result in a list of themes and questions about leadership for sustainability that can be used to help identify future work both in and, more importantly, outside of the classroom (~5 min).

Then, the instructor introduces the "World Café" process as a flexible format for hosting large group dialogue and harvesting a group's collective wisdom (~5 min). The first step is to form groups of four or five students and give each group a large sheet of paper (or flip chart). One student in each group should be identified as the "anchor," who will stay with the paper throughout the process and will record insights emerging from the conversation.

Each group begins by discussing two questions that address the connection between leadership and complexity (provided in ESM-B for projection):

1. What skills, mindsets, and dispositions do I need to cultivate in myself to more effectively engage with complex challenges and systems so that I can help create a more sustainable future?
2. What thought patterns, beliefs, or behaviors might I need to let go of to more effectively engage with complex challenges and systems so that I can help create a more sustainable future?

It is important to emphasize that the purpose of these conversations is not to establish consensus or "right answers." Rather, this process is designed to elicit multiple perspectives from divergent thinking and varied experiences. Anchors should record the group's insights. This can involve some artistic or graphic representation but should be descriptive and evocative so others who view the paper can understand the main points. Because the group is engaging with two related but different questions, it may be helpful for the anchor to divide the paper into two sections for recording purposes.

Students are given 15 min for the first round of conversation. The instructor can float among the groups in case any questions emerge about the process or prompts. After the first conversation has ended, ask the students to switch groups(with the anchor staying put) and form new ones with entirely different combinations of members (if possible, with none of the same people from their first group in the second one). The second conversation starts with the anchor giving a very brief synopsis of the previous group's conversation, while the new members make connections with the ideas from their first group's conversation. The second group then continues discussing the same guiding questions for an additional 10–15 min. The second round of conversation offers a chance for students to hear additional perspectives and explore these questions in greater depth. The anchor continues to record group members' insights.

After the second conversation, the anchors are asked to hang the papers around the room. The instructor hands each student a blank note card and invites the students to participate in a 5–10-min silent gallery walk, during which they should review and reflect on the recorded insights. The instructor asks each student to record three themes, insights, or questions that they think are most essential to reflect on or learn more about in relation to their own growth as a leader. The activity concludes with a large group conversation and share out about the exercise and the meta-level insights that emerged from the activity (5–10 min). Students should be encouraged to notice themes as well as divergent responses. The instructor can conclude the activity by asking students to consider the link between *understanding the nature of complex systems* and our ability to *cultivate leadership practices* that allow us to effectively engage with sustainability challenges.

Follow-Up Engagement

- Ask students to turn in their note cards to the instructor who can organize, synthesize, and share the responses back with the whole group to enhance the shared learning process. This information can be used to create a student-driven agenda/schedule for important topics to explore as the course progresses.
- Ask students to choose one of the topics/themes arising from the discussions and explore it in more depth through reflective writing and/or by exploring additional resources. Ask them to consider how that theme could be actively implemented in their personal and professional lives.
- As the course unfolds, provide additional opportunities for students to reflect on this activity and the insights it generated. Ask them what new insights and issues have emerged from other class topics and activities.
- Have students identify a complex sustainability challenge/case study that interests them and critically examine the dominant leadership practices/strategies that are being used to address the challenge. Ask them to consider how those practices might be shifted to more effectively engage with the complexity of the situation.

Connections

- Most sustainability-oriented issues whether related to ecology, social justice, economic viability, or other domains are complex challenges. This framework can help students recognize important characteristics of these challenges and begin to more confidently navigate the ambiguity associated with them.
- Cultivating leadership skills, mindsets, dispositions, and behaviors is helpful in empowering students to actively engage with almost any sustainability challenge (rather than feel overwhelmed by their enormity; e.g., see Chaps. 6, 8, 29, 39, and 40 of this volume).
- This activity can offer a relevant introduction to the field of systems thinking; helping students see how recognizing patterns and interrelationships can enhance our ability to *understand* and *address* sustainability challenges (e.g., see Chaps. 3, 28, and 31 of this volume).

References

Brown J (2005) The world café: shaping our futures through conversations that matter. Berrett-Koehler, San Francisco, CA

Ladyman J, Lambert J, Wiesner K (2013) What is a complex system? Eur J Philos Sci 3:33–67

Meadows D (2008) Thinking in systems: a primer. Chelsea Green, White River Junction

Rickles D, Hawe P, Shiell A (2007) A simple guide to chaos and complexity. J Epidemiol Community Health 61:933–937

Scharmer C (2009) Theory U: learning from the future as it emerges. Berrett-Koehler, San Francisco, CA

Senge P (2008) The necessary revolution: how individuals and organizations are working together to create a sustainable world. Doubleday, New York, NY

Smith M (2015) Three qualities of leadership that can catalyze change. http://www.uvm.edu/rsenr/leadership-sustainability/blog/18/three-qualities-leadership-can-catalyze-change. Accessed 15 June 2015

Snowden D, Boone M (2007) A leader's framework for decision making. Harv Bus Rev 85. https://hbr.org/2007/11/a-leaders-framework-for-decision-making. Accessed 15 June 2015

Wheatley M (2010) Finding our way: leadership for an uncertain time. Berrett-Koehler, San Francisco, CA

The World Café Community Foundation (2015) World Café methods. http://www.theworldcafe.com/method.html. Accessed 15 April 2015

Chapter 8
Discovering Authentic Hope: Helping Students Reflect on Learning and Living with Purpose

Kimberly F. Langmaid

> *Hope, authentic hope, can be found only in our capacity to*
> *discern the truth about our situation and ourselves and summon*
> *the fortitude to act accordingly.*
>
> David Orr (2011; p. 332)

Introduction

Learning about environmental and sustainability issues such as climate change, forest dieback, food and water shortages, the extinction crisis, and ocean pollution can become a source of acute and chronic stress and create feelings of disempowerment and fatalism (Clayton and Myers 2009). As we teach students about these challenges, educators can play a powerful role in facilitating affective learning experiences with outcomes that foster students' personal reflections about values, attitudes, and behaviors (Shepard 2008) and help nurture in them a positive vision of the future (Iliško et al. 2014) and a sense of purpose (Damon 2008). Tilbury (2011) suggests that education for sustainable development requires new pedagogical strategies and learning processes including "learning to clarify one's own values" and "learning to envision more positive and sustainable futures" to participate in creating a more sustainable and hopeful world. Orr (2011) makes a useful distinction between hope and optimism: "*Hope* is a verb with its sleeves rolled up. Hopeful people are actively engaged in defying the odds or changing the odds. Optimism leans back, puts its feet up, and wears a confident look, knowing that the deck is stacked" (p. 324).

Environmental and sustainability educators are called on to incorporate teaching strategies that nurture a psychology of hope (Clayton and Myers 2009). Authentic hope—the kind of hope which is based on personal values, talents, positive visions of

Electronic supplementary materials: The online version of this chapter (doi:10.1007/978-3-319-28543-6_8) contains supplementary material, which is available to authorized users.

K.F. Langmaid (✉)
Sustainability Studies Program, Colorado Mountain College, Edwards, CO, USA

Walking Mountains Science Center, Avon, CO, USA
e-mail: kiml@walkingmountains.org

© Springer International Publishing Switzerland 2016
L.B. Byrne (ed.), *Learner-Centered Teaching Activities for Environmental and Sustainability Studies*, DOI 10.1007/978-3-319-28543-6_8

Fig. 8.1 This *spiral diagram* conveys the ongoing, recursive personal-discovery process of authentic hope. It can be used to help students visualize and understand the significance of personal values, talents, vision, and purpose in creating a sense of authentic hope in their lives.

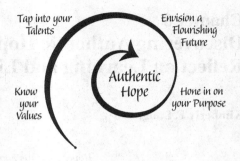

the future, and sense purpose—can encourage feelings of agency and foster pro-environmental behaviors (Ojala 2012). As students become more conscious of their values and take more opportunities to apply and practice their talents, they may be more likely to formulate a positive vision for their future and organize their lives in more purposeful and authentically hopeful ways (e.g., Fig. 8.1). For young people, the ongoing process of discovering and defining their purpose can foster greater hope, well-being, and life satisfaction (Burrow et al. 2010, Damon 2008). The activity in this chapter provides a reflective learning process to help students discover authentic hope in their lives by encouraging them to reflect on their personal values, identify their talents, articulate a positive vision of the future, write a brief personal purpose statement, and work together in small groups to explore their discoveries. Because authentic hope is not a static end point but is a state of being empowered in the process of creating a better future, students can be encouraged to revisit the topics of this activity throughout their lives for continued reflection on and refinement of their perspectives.

Learning Outcomes

After completing this activity, students should be able to:

- Reflect on and explain their personal values and talents.
- Write a vision statement of themselves living in a flourishing sustainable society.
- Synthesize their values, talents, and vision in a personal purpose statement.
- Recognize how a sense of purpose can lead to more hope and well-being in one's life.

Course Context

- Developed for a sustainability leadership course for majors and with ~25 students in a college-level sustainability studies program
- 90 min in two class meetings

- Requires some introductory knowledge of environmental and sustainability issues
- Adaptable for any class size

Instructor Preparation and Materials

The full implementation of the activity includes three components: (1) a 40-min pre-homework session facilitated by the instructor, (2) a worksheet homework assignment, and (3) a 50-min workshop-style class session. Instructors should be familiar with the following concepts in the context of sustainability issues: values, talents, vision, purpose, and hope (definitions are provided in Electronic Supplementary Materials—ESM-A). For additional background on the importance of affective learning outcomes in environmental and sustainability education, see Shepard (2008).

Prior to assigning the homework, the instructor should be prepared to facilitate a 40-min conversation and brainstorming session to review some of the environmental and sustainability issues and challenges students may have already studied in their courses (e.g., climate change, pollution, the biodiversity crisis) and help them connect those issues to the ideas of values, talents, vision, purpose, and hope with stories of actual people who have taken positive actions (Fig. 8.1 can be drawn or projected as visual aide for this discussion). The "Champions Chart for Instructors" (ESM-B), which asks students to provide examples they know of or can find online, can be used to facilitate this session. (For background and additional talking points, environmental educator and researcher Bardwell (1991) describes the importance of sharing environmental success stories and positive imagery with students.) As desired, to ensure some success stories can be discussed, the instructor can prepare examples to share (or assign students a preparatory reading) from Loeb (1999), Suzuki and Dressel (2002), and www.goldmanprize.organd http://wwf.panda.org/wwf_news/successes/.

For the homework assignment, the "Student Worksheet" (ESM-C) should be reviewed, edited as desired, and provided to each student (printed or electronically). Its purpose is to guide students through a self-reflective process of selecting their top five personal values, identifying their most important personal talents, writing a vision statement of a flourishing future, and writing a brief personal statement of purpose. Before assigning it, the instructor should complete the worksheet for herself to understand the process of the activity and share examples from her own worksheet if appropriate.

For the workshop class session, instructors should plan to organize students into small groups of four or five and adapt the "Conversation Guide and Gallery Walk for Students" sheet (ESM-D) to organize the session. Because students will share their homework reflections and collaborate to create posters using the "Authentic Hope Spiral Diagram" (Fig. 8.1), a large piece of paper (e.g., poster board size) and markers or crayons should be gathered for each group.

Activities

Pre-homework Session

1. Introduce the session (~5 min) by explaining that environmental and sustainability studies can be emotionally challenging; however, it is important to consider the positive aspects and opportunities for creating solutions.
2. Generate a chart on a chalkboard/whiteboard (e.g., ESM-B) and facilitate a brainstorming session (~15 min) to fill in the chart by asking students to share some examples of people who may have overcome emotional challenges and created positive solutions, who can be described as "environmental and sustainability champions." Once the stories of three to five champions have been shared, the instructor explains that these people possessed a sense of hope and empowerment that was likely an outcome of knowing their values, tapping into their talents, holding a positive vision of the future, and honing in on their sense of purpose.
3. Draw (or project) the "Authentic Hope Spiral Diagram" (Fig. 8.1) on a board and provide a brief overview (~10 min) on the concepts of values, talents, vision, purpose, and Authentic hope (see ESM-A).
4. Facilitate a deeper conversation (~10 min) about sustainability champions by asking students to brainstorm which values, talents, visions, and purposes each of the champions may have had. The list of values and talents from ESM-C can be used.
5. Tell students they will have the opportunity to work on discovering their sense of purpose by completing the "Student Worksheet" (ESM-C) for homework. Students should be asked to come prepared to share some of their reflections to the worksheet prompts during the next class session. Tell them that they do not have to have "perfect" answers but they should be reflective, honest, and meaningful.

Post-homework Session

1. Begin the session by telling students they will work in small groups to discuss their reflections from the homework worksheet and create a poster using the spiral diagram (Fig. 8.1).
2. Organize students into groups of four or five and provide each with the conversation guide (ESM-D) and a large sheet of paper and markers to create a group poster to share with the whole class.
3. Students talk together (~20 min) and take notes while sharing their worksheet reflections and providing answers to the four questions on the conservation guide. These questions are meant to help students make connections between their values, talents, and visions and help them articulate their sense of purpose.

4. Then they work together (~15 min) to create a collaborative poster using the spiral diagram (Fig. 8.1) and adding their notes or drawings on the poster. The goal is to create a final poster that is a visual reflection of the group's collective values, talents, visions, and purposes.
5. Students hang their posters around the room and then each group goes on a "gallery walk" (~10 min) to view, take notes, and discuss the collection of posters. Discussion questions to guide the gallery walk are provided in ESM-D; tell students they will be asked to share their responses verbally during a concluding discussion.
6. Close the session with a debriefing conversation about what students learned about themselves and their peers in relation to maintaining a sense of purpose and authentic hope for environmental and sustainability issues. Tell students their personal purpose statement can be an important source of inspiration and empowerment to keep them on track in their future aspirations and decisions. Encourage students to display their purpose statement in places they see on a daily basis (e.g., notebook, computer, or bathroom mirror) as a reminder to focus on a positive outlook and direction, to share them with their family and friends, and to revisit and revise it annually as it will likely evolve over their lifetimes.

Follow-Up Engagement

- Possible questions to guide extended discussion or that could be used to create writing or other assignments include:
  - How do your values help guide your decisions and behaviors when it comes to environmental and sustainability issues?
  - What personal traits might prevent you from taking action or hinder your ability to work toward a more sustainable life situation?
  - Is it possible that a person's sense of purpose can inadvertently work to further degrade the environment or create unsustainable situations? Think of some examples and share.
- Assign students to do a research and biographical writing project on a person who has created positive change for the environment or sustainability. Ask them to uncover what that person's values, talents, vision, and sense of purpose might have been.
- Ask students to reflect on their personal traits that hinder sustainability (internal barriers) or that they want to change in their life situations that interfere with the positive work they want to do. (For additional information on the concept of internal and external barriers to sustainable behaviors, see McKenzie-Mohr and Smith (1999).)
- Have students watch exemplary videos that share environmental and sustainability success stories and inspire positive visions and action, such as the talk by Paul Stamets (2008), entitled "6 Ways Mushrooms Can Save the World."

Connections

- Throughout any course on the environment and sustainability, instructors can try to balance negative with positive information and provide opportunities for students to learn about and envision positive futures.
- Ask students to reflect on values and how environmental and sustainability behaviors and issues are related to both individual and societal values. For example, which values and behaviors may contribute to climate change? Species extinction and the loss of biodiversity? Ocean pollution? Shortages of food and water? (For related activities, see Chaps. 4, 6, 29, and 39 of this volume.)
- Artistic expression can often foster inspiration and hope (e.g., see the activity in Chap. 43 of this volume).

References

Bardwell L (1991) Success stories: imagery by example. J Environ Educ 23:5–10

Burrow AL, O'Dell A, Hill P (2010) Profiles of a developmental asset: Youth purpose as a context for hope and well-being. J Youth Adolesc 39:1265–1273

Clayton S, Myers G (2009) Conservation psychology: understanding and promoting human care for nature. Wiley-Blackwell, West Sussex, UK

Damon W (2008) The path to purpose: how young people find their calling in life. Simon and Schuster, New York, NY

Iliško D, Skrinda A, Mičule I (2014) Envisioning the future: bachelor's and master's degree students' perspectives. J of Teacher Education for Sustainability 16:88–102

Loeb P (1999) Soul of a citizen: living with conviction in a cynical time. St. Martins, New York, NY

McKenzie-Mohr D, Smith W (1999) Fostering sustainable behavior: an introduction to community-based social marketing. New Society, Gabriola Island, BC

Ojala M (2012) Hope and climate change: the importance of hope for environmental engagement among young people. Environ Educ Res 18:625–642

Orr D (2011) Hope is an imperative: the essential David Orr. Island Press, Washington, DC

Shepard K (2008) Higher education for sustainability: seeking affective learning outcomes. Int J Sustain High Educ 9:87–98

Stamets P (2008) 6 ways mushrooms can save the world. http://www.ted.com/talks/paul_stamets_on_6_ways_mushrooms_can_save_the_world?language=en. Accessed 24 July 2015

Suzuki D, Dressel H (2002) Good news for change: how everyday people are helping the planet. Greystone, Vancouver, BC

Tilbury D (2011) Education for sustainable development: an expert review on process and learning. UNESCO, Paris, France

Chapter 9
Teaching How Scientific Consensus Is Developed Through Simplified Meta-analysis of Peer-Reviewed Literature

Emily S.J. Rauschert

Introduction

All students, especially those studying environmental and sustainability issues, should understand how the process of science works and how scientists deal with conflicting results among studies (Bryce and Day 2014). However, teaching them about the diverse pathways to scientific knowledge and how scientists develop consensus about emerging questions is challenging. Many students are not skilled readers of the peer-reviewed scientific literature and have trouble synthesizing across articles. Nonetheless, most faculty agree that the primary literature should be used when teaching undergraduate students (Coil et al. 2010), making it necessary to provide them with opportunities to engage with the literature in a variety of ways. In particular, asking students to interpret and synthesize studies that reach different conclusions about a topic (e.g., global environmental change) can help them become more skilled thinkers and engage them in discussions and analyses that mirror how professional scientists work to develop scientific knowledge and consensus (also see Chaps. 15 and 34, this volume).

Meta-analysis is a statistical method used to evaluate and synthesize a set of studies on a topic to gain deeper insights and reach more reliable conclusions, especially when studies reach divergent conclusions. It has been increasingly used in ecology and environmental science over the past 20 years (Gurevitch et al. 2001). To increase students' abilities to synthesize many studies on a topic and improve their skills for reading peer-reviewed literature, the activity in this chapter

Electronic supplementary materials: The online version of this chapter (doi:10.1007/978-3-319-28543-6_9) contains supplementary material, which is available to authorized users.

E.S.J. Rauschert (✉)
Department of Biological, Geological and Environmental Sciences, Cleveland State University, Cleveland, OH, USA
e-mail: e.rauschert@csuohio.edu

L.B. Byrne (ed.), *Learner-Centered Teaching Activities for Environmental and Sustainability Studies*, DOI 10.1007/978-3-319-28543-6_9

involves conducting an informal and simplified meta-analysis using a set of articles related to a common topic. Its primary goal is to build students' scientific thinking skills, including quantitative reasoning, and help them better understand how scientific consensus is developed, especially for topics that may be perceived as controversial by the public. These skills have been identified as core competencies crucial for undergraduate biology education (Brewer and Smith 2011). The activity can be used with a wide variety environmental and sustainability studies topics; an example case study about the effects of global change on invasive plant species, with a prepared list of articles and example results, is shared for quick and easy adoption.

Learning Outcomes

After completing this activity, students should be able to:

- Interpret and explain graphs in and extract major conclusions from a scientific peer-reviewed article.
- Identify the procedure of a meta-analysis and conduct a simplified meta-analysis.
- Synthesize results of multiple studies to reach a general conclusion for a question, identifying both the overall consensus and any contradictions.

Course Context

- Developed for an upper-level majors course in invasion ecology for science majors with 20 students
- 50 min in one class meeting
- Before class, students should have some prior experience with graph interpretation and reading primary literature articles and be familiar with different types of scientific studies (experiments, observational studies, models, reviews). The activity requires previous background knowledge about the chosen topic investigated to complete the meta-analysis activity. For example, in the original context, students had some limited prior knowledge of the major components of global change, covered in a previous class
- Adaptable for course sizes of up to 60 students and for a variety of topics where some contradiction or confusion about results and conclusions exists in the literature. It was originally implemented to examine the effect of global change on invasive species, but it is suitable for other topics, such as the benefits of conventional versus organic agriculture, the ecosystem effects of invasive earthworms or impacts of genetically modified organisms

Instructor Preparation and Materials

Instructors should be familiar with meta-analysis methods and prepare an introduction for students; for background, see Box 9.1, Gurevitch et al. (1992, 2001), Arnqvist and Wooster (1995), and Borenstein et al. (2009). In addition, prepare for a brief discussion on why ecological studies do not always lead to exactly the same results (e.g., spatial and temporal abiotic variation, stochasticity), which will be used to start the class session; suggested questions are provided in section "Activities."

For the activity, a topic should be selected that is relevant to the course and contains some divergent or contradictory results in the literature. Identify a clear question that is to be answered by the meta-analysis and will guide the student's work; the topic and question should be focused and coherent but could include multiple variables (e.g., in the invasive species case study, four variables were examined; see

Box 9.1. A Brief Introduction to Meta-analysis

Meta-antalysis is a quantitative way to statistically synthesize information across multiple studies. In general, a meta-analysis proceeds as follows. A hypothesis is developed, such as "An increase in CO_2 will positively influence invasive plants." Typically, a search criterion is developed, such as "we want to include all papers on the topic of invasive plants and CO_2 increases, published in the last 20 years, involving experimentation." Using an online search engine, a large list of candidate papers will be generated, but the articles should be screened for suitability; many might be excluded from the analysis because they don't meet the preset criteria (e.g., not reporting experiments) or they don't relate clearly to the hypothesis being investigated. All suitable papers are examined, and certain predefined information is extracted from them such as whether a positive or negative impact is seen.

Typically, an attempt is made to calculate something from the results of each study; one common metric is effect size, which is often a standardized difference in the means of experimental and control groups or the response ratio, which is the ratio of a measurement in experimental and control groups (Gurevitch et al. 2001). When analyzing all of the studies, the effect sizes or response ratios can be weighted by the sample size of the studies or the standard error of the effect size to give more weight to larger studies. It is then possible to statistically draw inference about whether a particular hypothesis is supported across the literature. In the simplified approach described for this activity, students tally the direction of the effects in a vote-counting step, i.e., positive, negative, and no effect (denoted in the matrix by +, −, and 0, respectively). Note that this vote-counting approach is a simplified first step and that formal meta-analysis would require the calculation of effect sizes or similar metrics (Koricheva and Gurevitch 2014).

Box 9.2. Case Study Examining the Effects of Global Change on Invasive Plants

The impacts of four major components of global change on invasive plant success were identified (increasing temperature, increasing CO_2, nitrogen deposition, and changes in precipitation); the papers listed in ESM-A relate to the effects of one or more of these components on plant invasion. If the class size is less than ten, it may be best to limit the focus to fewer components so that multiple papers are used about each aspect of global change investigated. It might be good to do an additional literature search to identify very recent papers on this topic as well.

To prepare students for this activity, during the class before the activity, introduce the four main aspects of global change if they are not already familiar with them. If more than one aspect of global change is being considered, make a table on the board to prepare for the results (see Table 9.1 and Fig. 1 in ESM-A). For each paper, students will then later add to the table whether the paper showed a positive, negative, or neutral effect on plant invasion. Print out enough copies of the in-class worksheet (see ESM-B), so that students can better organize their assessment of the papers. When originally implemented, the students reached conclusions similar to the published meta-analysis on this topic (Bradley et al. 2010).

Box 9.2). The students will be collecting information about whether a hypothesis is supported in each paper or not (e.g., in the case study, whether aspects of global change will increase plant invasion); prepare a table where students will compile this information (see Table 9.1 and Fig. 1 in Electronic Supplementary Materials ESM-A for an example).

For the topic, identify a set of candidate papers that could be assigned for students to read by conducting an online search. Ideally, the topic should have at least 30 papers published about it so that there are enough to choose from. To narrow the search results, consider time limits where appropriate, such as all papers used in the past 20 years, if, for example, the topic is best examined using more recent techniques. Think about the type(s) of study to include and exclude (e.g., experiments, observational studies, or mathematical models). It is best to focus on papers that present novel data and avoid including multiple papers that use the same data. It should be possible for students to readily extract information from the paper from graphical presentations of the results. Exclude literature reviews, and consider avoiding papers that seemed to have experimental designs or results that are likely to confuse students. If there is a published full meta-analysis about the topic, its results can be used to compare with the class results to foster additional discussions. The published meta-analysis could also serve as a source of papers to use for the activity. For the activity, select and print out copies of enough papers for the class size (at least one different per student and more if time permits). Alternatively, students could be provided with the digital copies to read on electronic devices in class or for

Table 9.1 Sample simplified meta-analysis results table

	Effects of global change on invasives			
	Increasing temperature	Increasing CO_2	Changes in precipitation	Nutrient deposition
Individual results	+	+	+/−	+
	−	+	−	+
	+	+	+	−
	+	0	+	+
	−	+	−	+
				0
Final result	+/−	+	+/−	+

This table was generated using some of the papers listed in Electronic Supplementary Materials A

homework; the latter may be important if there is a concern with less experienced students not being able to read articles quickly enough in class. Save electronic copies of all of the papers on the computer used for classroom projection, so that the graphs students wish to present can be projected.

Students will work in pairs to analyze two articles, and it may be useful to pair up stronger and weaker students in advance. As desired, prepare a worksheet (and print one for each student) to guide students reading and assessment of the papers; an example for the invasive species case study that can be adapted for other topics is presented in ESM-B.

To implement the activity with advanced options, give the students some papers that are not quite suitable for the meta-analysis and allow them to participate in the screening and selection process. In addition or alternatively, have students conduct the literature search themselves, either out of class as a homework exercise or with guidance in class. It would be possible to extend this activity and have students actually calculate effect sizes, but this would require more than one class period.

Activities

Begin the class with a discussion about variability that helps students understand why conflicting results for similar studies might be expected. Example questions and talking points include:

- Do you think the same experiment conducted in different places will always give the same results?

  - Ecological systems are highly complex, and variation from day to day and year to year is common; thus, experiments conducted at different times or places are likely to have different outcomes.

- It is not uncommon for studies to show different results, so how can scientists generalize from studies that don't always agree?

  – Introduce meta-analyses as a rigorous, quantitative method that scientists use to synthesize results from many studies for a topic, especially ones that have variable results. Additional information can be presented from Box 9.1 and the references cited above.

After introducing the activity (and, as needed, the focal topic and question for the meta-analysis), assign two or three articles per student pair. Ask students to extract summary information from each paper (15–20 min.) that helps answer the question. For the meta-analysis, each student pair needs to conclude whether the results had a positive, negative, mixed, or neutral impact on the (dependent) variable of interest (e.g., in the case study, whether the plant invasion was increased, decreased, or not affected by the global change variable). A prepared worksheet that students complete can help focus their attention on relevant information and expedite this step (see the example for the case study in ESM-B). As part of the introduction, the instructor should direct students' attention to the meta-analysis matrix (table), written on a board or other surface, into which students will record the results of the papers.

Next, each pair should present their papers by projecting and explaining relevant figures that demonstrate the main results (~2–5 min per pair, depending on class size). For this, the instructor should have the papers readily accessible for quick projection. However, it is recommended to let the students identify which figure(s) should be projected for a more learner-centered approach; this is a useful check on whether the students' interpretations of graphs are valid. Based on the study's results and conclusions, students then make a +, −, +/−, or 0 in the meta-analysis table (e.g., Table 9.1 and Fig. 1 in ESM-A). If the class size is too large to allow all students to present, it may be better to have some students present while ensuring that all pairs add their results to the table.

Examine all results in the table as a class and help guide students, via discussion, to an overall conclusion based on the numbers of positive, negative, and neutral results from across all the examined papers. Although there will likely not be 100 % consensus on the direction of the effects, students can conclude what type of effect was observed by most studies. If there is a published meta-analysis on the topic, the class results can be compared to its results (e.g., Bradley et al. 2010 for the case study). In the case study example, even though the simplified approach was coarser and less quantitative than the published meta-analysis, similar conclusions were reached.

To end the activity, these questions can be used in a final discussion:

- What can we conclude about the topic investigated? (For the case study, what are the possible impacts of global changes on invasive plants?)
- Were we able to reach a reasonable consensus? What would strengthen our ability to draw valid conclusions? What was left out by this approach that might be important? (Discuss the value of more studies, larger sample sizes, data from many ecosystems, etc.)

- Should experimental, observational, and modeling studies be given equal weight in such an analysis?
- How representative were the papers examined, in terms of where they were conducted, which species were used, etc.?
- How does the variation in results across studies affect our ability to draw strong conclusions or make predictions?
- How could our approach be improved? (Discuss effect sizes, weighing results with the sample size of the study, and point out that actual meta-analyses are more quantitative than just a +/− approach.)
- How easy do you think it is to publish positive, negative, and neutral results? How will this affect our ability to draw valid conclusions? (Discuss the file drawer problem and how publication bias might influence the conclusions reached by meta-analysis.)
- Does communicating some level of uncertainty undermine public confidence in scientific results? Under what circumstances is it important to communicate uncertainty?

Follow-Up Engagement

- Have students go through the same exercise with a different topic and write a report about their results.
- Discuss how scientific conclusions and consensus may be used to guide policy- and decision-making. How should policy be designed around scientific conclusions in the light of conflicting research outcomes? What roles do meta-analyses have for guiding policy-making and other decisions?
- On an exam, give students a hypothesis and ask them to come up with nonexperimental ways to test it (i.e., to assess if they can recall and describe a meta-analysis approach).

Connections

- This activity can be used to examine many topics in environmental sciences and sustainability, particularly where there is difficulty reaching consensus such as organic versus conventional agriculture, the impacts of genetically modified organisms, or whether biocontrol is beneficial.
- Throughout a course, encourage students to keep thinking about the different ways in which we gain scientific knowledge. Instructors commonly have students design experiments to address hypotheses; as a complementary approach, provide opportunities for which students might propose a meta-analysis to investigate an environmental question (e.g., see Chaps. 15, 32, 33, and 34, this volume).

References

Arnqvist G, Wooster D (1995) Meta-analysis: synthesizing research findings in ecology and evolution. Trends Ecol Evol 10:236–240

Borenstein M, Hedges LV, Higgins JPT, Rothstein HR (2009) Introduction to Meta-analysis. Wiley and Sons. http://www.wiley.com/WileyCDA/WileyTitle/productCd-EHEP002313.html. Accessed Jan 2015

Bradley BA, Blumenthal DM, Wilcove DS, Ziska LH (2010) Predicting plant invasions in an era of global change. Trends Ecol Evol 25:310–318

Brewer CA, Smith D (2011) Vision and change in undergraduate biology education: a call to action. Am Assoc Adv Sci

Bryce TGK, Day SP (2014) Scepticism and doubt in science and science education: the complexity of global warming as a socio-scientific issue. Cult Stud Sci Educ 9:599–632

Coil D, Wenderoth MP, Cunningham M, Dirks C (2010) Teaching the process of science: faculty perceptions and an effective methodology. CBE Life Sci Educ 9:524–535

Gurevitch J, Morrow LL, Wallace A, Walsh JS (1992) A meta analysis of competition in field experiments. Am Nat 140:539–572

Gurevitch J, Curtis PS, Jones MH (2001) Meta-analysis in ecology. Adv Ecol Res 32:199–247

Koricheva J, Gurevitch J (2014) Uses and misuses of meta-analysis in plant ecology. J Ecol 102:828–844

Part II
Ecology, Ecosystems and Environmental Management

Chapter 10
Understanding Ecosystems and Their Services Through *Apollo 13* and Bottle Models

Carmen T. Agouridis and Tyler M. Sanderson

Introduction

The concepts of ecosystems, their complexity (e.g., interacting biotic and abiotic elements), and the goods and services they provide are important to understand when examining environmental and sustainability topics (Pickett and Cadenasso 2002, Cadenasso et al. 2006). Earth's diverse ecosystems range in size, climate, and composition. Collectively, they provide numerous life-sustaining goods and services such as food, air purification, and pollination (Kremen 2005), yet their value is generally underestimated in part because it is difficult to place monetary values on such intangible items and processes (Costanza et al. 1997). Many people often assume that ecosystems are reliable and resilient and that their goods and services are thus unlimited; however, this assumption is not true. As humans place increasing pressures on ecosystems, they can be degraded and result in issues including water shortages, pest outbreaks, habitat loss, and species extinction. As such, society will benefit from better understanding ecosystem dynamics so that resources can be managed more sustainably to prevent their degradation (also see Chaps. 11–16, this volume).

Recognizing that ecosystem components are interconnected (i.e., a change in one component directly or indirectly affects others) is an important step in understanding ecosystem dynamics. The goal of the learning activity described below is to help students define ecosystems, identify biotic and abiotic components of an ecosystem, identify the ecosystem goods and services valued by humans, and understand how ecosystem components affect each other. This will be accomplished with movie

Electronic supplementary materials: The online version of this chapter (doi:10.1007/978-3-319-28543-6_10) contains supplementary material, which is available to authorized users.

C.T. Agouridis (✉) • T.M. Sanderson
Department of Biosystems and Agricultural Engineering, University of Kentucky, Lexington, KY, USA
e-mail: carmen.agouridis@uky.edu; tyler.sanderson@uky.edu

clips from *Apollo 13*, examining a model ecosystem and creating a visual web of an ecosystem's interconnectedness, all of which prompt a discussion of connections between biotic and abiotic elements and how ecosystems relate to human well-being. The knowledge gained from these activities should help students better understand how human actions can positively and negatively impact ecosystems.

Learning Outcomes

After completing this activity, students should be able to:

- Describe the concept of "ecosystem" and the range of ecosystems' sizes and compositions.
- Differentiate between abiotic and biotic components of an ecosystem and identify how they are interconnected.
- Describe ecosystem goods and services and explain their benefits for human well-being.
- Identify how the diversity of and interactions within ecosystems influence ecosystem goods and services.

Course Context

- Developed for a college first-year course with 20–25 students with interests in the environment and sustainability
- 100 min total in one or two class periods
- No background knowledge is needed
- Adaptable to courses of any level and class size (with modifications for larger or smaller group activities) that include ecosystem discussions and can be modified for longer or shorter durations by assigning activities outside class time and shortening discussion times

Instructor Preparation and Materials

The instructor should be prepared to (a) define "ecosystem" and describe different types of ecosystems, (b) define and provide examples of biotic and abiotic ecosystem components (e.g., water, soil, rocks, air, nutrients, plants, microbes, and animals), (c) discuss how an ecosystem's components are interrelated, (d) define and provide examples of ecosystem goods and services, and (e) discuss the monetary value of ecosystem goods and services. For (a)–(c), Nature Education (2014) and the linked pages therein provide basic definitions and a review of these concepts; unfamiliar readers could also consult with introductory biology and ecology textbooks for more background as needed. For (d) and (e), Table 1 from Costanza et al. (1997) lists 17 types of ecosystem goods and services (e.g., climate regulation, water supply, and pollination), summarizes their benefits for humans, and provides estimates of their monetary values (also see Costanza et al. (2014) for an update). Additionally, instructors can refer to Groffman et al. (2004) for further explanation and useful diagrams of ecosystems' structure and function.

The instructor should become familiar with the premise of the movie *Apollo 13* (1995), review and download five clips from it (12:10 min total time), and be able to discuss how each clip reflects the ecosystem concept (background information and links are provided in Electronic Supplementary Materials (ESM)-A). Instructors should also read Costanza et al. (1997); Table 1 will be used in Part 3 of the activity and can either be projected on a screen or given to students as a handout. If desired, an ecosystem worksheet (ESM-B) for student use during the learning activity could be downloaded and printed.

Prior to the start of class, the instructor should also choose a model ecosystem for students to examine (see Part 4 below). The model can be (1) an *Ecosystem in a Bottle* (described below), (2) an image (to be projected or provided as a handout; an example is provided in ESM-C), or (3) a nearby outdoor location (visible from the classroom or that the students are taken to). The chosen model should have as many components as students in the class to allow for individual student representation of each ecosystem component in the interconnectedness activity as outlined in Part 5 below. If the class size is larger than 25 students, consider splitting the class into two groups for Part 5 as there may be more students than ecosystem components. If using the constructed *Ecosystem in a Bottle*, instructors must decide whether to construct one for each group (approximately three to five students per group) or one for the entire class.

In the authors' experiences, the bottle ecosystems are the most effective model because they allow for small group participation while in the classroom setting and are an intriguing hook and conversation topic. These small, enclosed ecosystems are made from two-liter soda bottles. Instructions for constructing the *Ecosystem in a Bottle* are available through Mitton (2010), Wisconsin Fast Plants (2003), and in ESM-D. Depending on what materials the instructor already has available, costs to construct an *Ecosystem in a Bottle* can range from US\$ 5 to US\$ 45. While it is not critical to use the exact species listed in ESM-D, inclusion of at least one of each of the following biotic organisms is recommended: goldfish, snail, water plant (oxygenator), insects (any type), and terrestrial plant(s) (e.g., grass). Abiotic items that should be included are rocks, soil, water, and decomposing organic matter (e.g., fruit, leaves, and twigs). A full list of ecosystem components is provided in slide 5 of ESM-D.

Instructors are advised to review the lesson activities and estimated times for completing each activity below before implementing the lesson. Instructors may wish to reduce the amount of in-class time allotted for the activities (e.g., have students watch the movie clips prior to class, decrease discussion time) to conduct all activities during a single class. Likewise, instructors may wish to conduct the activities over two class periods (e.g., Parts 1–3 during the first class period and Parts 4–5 during the second class period).

Additional materials needed to complete the activity include two packs of differently colored Post-It notes or labels (one color each for abiotic and biotic components; at least one label for each student), one marker, and one spool of string or yarn (~100 ft.) for each group that will be making a web (i.e., one group for a class with <25 students or two for >25 students). As an alternative, the instructor could also prepare photos of the ecosystem components to post on the board during Part 5 to create the ecosystem web (see ESM-C).

Activities

Part 1: Apollo 13 Movie Clips (15–20 min)

At the start of class students watch the five *Apollo 13* movie clips. The instructor could prompt students to think of ecosystems while watching the movie clips, but in the authors' experiences, not doing so creates a more engaging exercise because the mysterious nature of this approach tends to capture the students' attention. Instead, the instructor can ask students if they have seen the movie and, as desired, give a brief overview of the Apollo 13 mission. After viewing, the class should discuss what was happening in each clip in the context of the Apollo 13 mission. (In the authors' experience, discussions in Part 2 are more fruitful if this discussion is explicitly included.) At this point, it is recommended that further discussion of how the movie clips relate to ecosystems be delayed until after Part 2 is completed. Students can be told that the movie clips will be revisited following a brief discussion about ecosystems.

Part 2: Ecosystem Discussion (15–20 min)

Next, the instructor asks the class to define the term "ecosystem." To ensure engagement, the instructor could ask students to write responses individually or conduct a think–pair–share. Then the instructor should solicit responses and write them on the board before presenting a definition using the students' best responses; this definition should be based on level and ability of the class. Next, the instructor asks for examples of ecosystems. Students will likely name large familiar ones such as the Amazon rainforest but may need prompting to list smaller ecosystems such as ponds, gardens, and patches of trees. Once a set of diverse ecosystems are identified, the instructor highlights the breadth of their sizes, types, and climates. At this point, the instructor asks students to write down how the *Apollo 13* movie clips are related to the topic of ecosystems (see ESM-A for suggestions). The instructor should solicit responses and write them on the board, noting that the connections will be revisited later in the lesson.

Part 3: Ecosystem Goods and Services Discussion (15–20 min)

Following the ecosystem discussion, the instructor asks the class to define the more general terms "goods and services" (i.e., in an economic sense) and provide examples of each, with responses recorded on the board. (For instance, goods are physical items such as books and cell phones, while services are intangible actions and processes such as cooking dinner and getting a haircut.) The instructor then asks the class as a whole, individually, or as a think–pair–share to define "ecosystem goods and services" and provide examples of each; responses are written on the board. Depending on the level and ability of the class, the instructor may need to provide an initial example to help guide students. Next, the class compares these class lists to Table 1 from Costanza et al. (1997) which can be projected or given as a handout. The instructor could lead a short discussion about the table, beginning with any previously learned information (e.g., plants provide oxygen, shade, and food) and ending with some less well-known services. Next, the instructor introduces the concept of estimating global value of all the Earth's ecosystem goods and services and the difficulties and uncertainties in valuing intangibles (e.g., pollination). Students could be asked to guess this annual value before it is shared: US\$ 125 trillion (10^{12}) annually (Costanza et al. 2014). The instructor should ask the students what happens if an ecosystem good or service is reduced or lost, whether it could be replaced, and if so, how. The instructor should prompt the students to think back to the *Apollo 13* movie clips, asking, what service was lost and what did the astronauts do to try and replace it? If desired, the instructor may extend the discussion by asking students to evaluate solutions to replace lost ecosystem good and services in this context.

Part 4: Ecosystem in a Bottle (15–20 min)

As a transition, the instructor may wish to briefly review the main points covered during Parts 1–3. The instructor then divides the class into groups of three to five students and introduces (and distributes as relevant) the model ecosystem (i.e., the *Ecosystem in a Bottle*, image, or outdoor view). Each group is asked to first observe and then list the model's abiotic and biotic ecosystem components. Lists of suggested ecosystem components are included in ESM-C (image model) and ESM-D (bottle model). One by one, the instructor calls on students asking each to identify an ecosystem component and state whether it is abiotic or biotic. The name of the ecosystem component is written on a color-coded label (one color each for abiotic and biotic elements) and given to the student. Students place their labels on their shirts, taking the identity of that component, and standing in a designated spot in the room until all students have been given an ecosystem component. If there are more students than ecosystem components, the instructor should have prepared materials for two (or more) webs. (For each identified component, two labels will be created and given to two students; these students should stand in separate designated spots in the room creating two groups.) If additional components are needed, students could be asked to list components not found in the model(s).

Part 5: Ecosystem Component Interconnectedness (20–25 min)

With students in the roles of components from an ecosystem, the instructor leads them through the creation of an ecosystem web. The instructor selects one student to start the web (e.g., water, sun, or soil) and gives him/her the end of the string. The instructor asks that student to identify one component to which they are connected from among the labeled students and to explain the connection. The spool of string is unwound to connect the starting and identified components. The student identified by the starting student, then describes a connection to a third component, and physically establishes the relationship by passing the string to that student. This process is repeated until all ecosystem components (i.e., students) are connected via the string thus creating a web. The final student to be selected can identify a connection to one of the components already selected. Remind the students to hold the string tightly in their hands so the web remains intact as it is developed. Alternatively to using string and students, the instructor may choose to post photos or labels of the ecosystem components on the board and have the students draw lines to connect the components (see an example in slide 4 of ESM-C).

Once the web is completed, the instructor simulates a few disturbance or stress events. For example, a drought is simulated by pulling water away from the group (while all students including water still hold the string) thus causing the strings to tighten and pull on other ecosystem components. Other examples of stressors are species loss, pollution, competition, invasive species, and population growth. The instructor can use their discretion developing their own approaches to altering the web based on class context and interests. With each event, the instructor asks the class what happened when one ecosystem component was "disturbed": how were the other ecosystem components affected and how might this affect ecosystem goods and services? The instructor should encourage the students to relate the interconnectedness of ecosystem components back to the model ecosystem analysis and the *Apollo 13* movie clips. For example, what happens when the amount of heat (i.e., thermal energy as measured by temperature) inside the *Ecosystem in a Bottle* increases (e.g., climate change)? What would happen if a pollutant was inserted into the *Ecosystem in a Bottle?* What happened when one of the oxygen tanks was destroyed in *Apollo 13* and the astronauts were forced into the lunar module? The instructor should wrap up the activities by helping the students relate what they learned to a real ecosystem with regard to goods and services provided and interconnectedness of ecosystem components. Suggested responses to these questions and a wrap-up example are provided in ESM-E.

Follow-Up Engagement

- Additional questions and suggested ecosystems are provided in ESM-F.
- Assign students a short writing assignment in which they use Table 1 from Costanza et al. (1997) as a guide to discuss the human benefits from a specific ecosystem service that is important to them.
- Explore how humans can protect and restore ecosystems. Assign students an ecosystem (suggestions provided in ESM-F) and ask them to research and list (1) threats to the ecosystem, (2) how the ecosystem can be protected, and (3) effective ways to restore or reclaim existing damage. (See Chaps. 12, 13, 16, 17, 19, and 20, this volume, for related activities.)

Connections

- The lesson can be related to the loss of habitats and biodiversity and the politics of conservation. Consider the reasons for legislation such as the Clean Water Act, Clean Air Act, and Endangered Species Act, and the impact of such legislation on ecosystem health.
- Throughout a course, ask students to analyze a topic considering anthropogenic impacts on ecosystems (e.g., agriculture, deforestation, mining, and urbanization) and how those can alter the connections among ecosystem's components (e.g., Chaps. 11–20, this volume).
- The impact of economic-based decisions on ecosystems can be explored using an *Ecosystem in a Bottle*. (For a game-based economic activity, see Verutes and Rosenthal (2014).)
- The lesson can be used as a starting point to examine pollution and environmental justice. For example, the extraction and processing of natural resources often occurs in economically disadvantaged regions of the world resulting in higher levels of pollution which disproportionately affect local communities. (For related references and activities, see Chaps. 27–33, this volume.)

References

Cadenasso ML, Pickett STA, Grove JM (2006) Dimensions of ecosystem complexity: heterogeneity, connectivity, and history. Ecol Complex 3:1–12

Costanza R, d'Arge R, de Groot R et al (1997) The value of the world's ecosystem services and natural capital. Nature 387:253–260

Costanza R, de Groot R, Sutton P et al (2014) Changes in the global value of ecosystem services. Glob Environ Chang 26:152–158

Groffman PM, Driscoll CT, Likens GE et al (2004) Nor gloom of night: a new conceptual model for the Hubbard Brook Ecosystem Study. Bioscience 54:139–148

Kremen C (2005) Managing ecosystem services: what do we need to know about their ecology? Ecol Lett 8:468–479

Nature Education (2014) Ecosystem Ecology. http://www.nature.com/scitable/knowledge/ecosystem-ecology-13228212. Accessed 29 Apr 2015.

Pickett STA, Cadenasso ML (2002) Ecosystem as a multidimensional concept: meaning, model and metaphor. Ecosystems 5:1–10

Mitton M (2010) Kids' summer crafts: build an ecosystem [Web log post]. http://scribbit.blogspot.com/2010/05/kids-summer-crafts-build-ecosystem.html. Accessed 5 Jan 2015

Verutes GM, Rosenthal A (2014) Using simulation games to teach ecosystem service synergies and trade-offs. Environ Pract 16:194–204

Wisconsin Fast Plants Program (eds) (2003) Bottle biology, 2nd edn. Kendell Hunt, Dubuque

Chapter 11
Using Soil Organisms to Explore Ecosystem Functioning, Services, and Sustainability

Jessica Baum and Rachel Thiet

Introduction

Soil organisms play crucial roles in the maintenance of ecosystem functioning and the provision of ecosystem services, agricultural productivity and sustainability, ecological and community resilience, and climate regulation (Nardi 2007; Bardgett and Wardle 2010; Schimel 2013; Wagg et al. 2014). Despite increasing global awareness about these critical roles of soil organisms (Wall and Six 2015), many classes that include content about soil emphasize physical and chemical properties (e.g., pH, texture, and moisture) while neglecting soil biology. This is problematic because students cannot achieve deep understanding of ecosystems without knowledge about soil organisms and their ecology. This, in turn, compromises the effectiveness of our students when they become professionals in ecology, farming, teaching, public planning, resource management, and many other areas. Additionally, expanding the realm of student knowledge to include belowground organisms and activities enriches the understanding they bring into their personal lives as citizens of the world.

Further, courses that de-emphasize or ignore soil organisms miss an opportunity to engage students in fun, experiential learning about some of the most unique and fascinating organisms on the planet. The goal of the playful, learner-centered activity described in this chapter is to introduce students to soil organisms, their relationships to one another, and their roles in ecosystem functioning, ecosystem services, and sustainability. It is adapted from an activity from Byrne (2013) by adding a station-by-station, student-led teaching and worksheet component (possibly with

---

Electronic supplementary materials: The online version of this chapter (doi:10.1007/978-3-319-28543-6_11) contains supplementary material, which is available to authorized users.

J. Baum (✉) • R. Thiet
Department of Environmental Studies, Antioch University New England, Keene, NH, USA
e-mail: jbaum@antioch.edu; rthiet@antioch.edu

© Springer International Publishing Switzerland 2016   97
L.B. Byrne (ed.), *Learner-Centered Teaching Activities for Environmental and Sustainability Studies*, DOI 10.1007/978-3-319-28543-6_11

live organisms) to emphasize the connections of soil organisms to ecosystem functioning and services, sustainability, and human well-being. Through engaging in active inquiry and discovering soil creatures and their ecological functions, students can begin to understand the world beneath their feet and appreciate that soil organisms are essential for maintaining ecosystem sustainability and human well-being (also see Chap. 10, this volume).

Learning Outcomes

After completing this activity, students should be able to:

- Identify and describe various soil organisms, their relationships in food webs, their ecological functions in soil, and the benefits they provide to human society.
- Explain the primary ecosystem functions and services of one taxonomic group of soil organisms.
- Apply new understanding of soil ecology and soil food webs to their analysis of ecosystem services in various ecosystem types, including agricultural ones.
- Integrate understanding of soil organisms into broader comprehension of ecology, ecosystem functioning and services, sustainability, and human well-being.

Course Context

- Developed for a graduate-level soil ecology course with 15–20 students
- 60 min in one class meeting, with 5–10 min preparation in the previous class
- Before class, students research and prepare to present about their assigned organisms. Ideally, students will have had some instruction in taxonomic biodiversity and ecosystem ecology
- Adaptable to various class sizes (e.g., 10–40 students) and learning levels (e.g., high school, college, and graduate school). The activity can be introduced into any ecology, biology, environmental science/studies, or sustainability course at various points in the curriculum

Instructor Preparation and Materials

Prior to this activity, instructors should review the primary soil food web constituents, their relationships, and their main ecological functions (see Electronic Supplementary Materials (ESM)-A for suggested resources). Also, the instructor must prepare assignment cards to distribute to students the week prior to the in-class

Box 11.1. Sample Soil Organism Groups

- Soil bacteria (e.g., actinomycetes, azobacteria)
- Mycorrhizal fungi
- Saprophytic fungi (e.g., basidiomycetes)
- Nematodes
- Protozoa (e.g., amoeba, ciliates, euglena)
- Mites
- Springtails
- Earthworms
- Enchytraeids (white worms)
- Insects (e.g., ants, beetles)
- Arachnids

activity. These $3 \times 5''$ cards should have the names of various groups of soil organisms (hereafter "organisms") at the top of each (see Box 11.1 and also Byrne 2013). The instructor should select half as many organisms as there are students and prepare two cards per organism (e.g., for a class of 20 students, 10 organisms would be selected with two cards per organism). The instructor randomly hands out one of the organism cards to each student. Students are then prompted to engage in an internet scavenger hunt for large color images and/or videos (e.g., from YouTube) and information about their organisms. Students are to become "experts" on their assigned organisms so they can teach their classmates about them. To prepare for this, they should be told to bring several printed or electronic images or videos (along with electronic devices to show the latter two) of their organisms to class on the designated date and, based on their research, write the following information on their cards (modified from Byrne 2013):

1. What their organisms eat
2. What eats their organisms
3. What their organisms do in soil (their functions) and their effects on ecosystems
4. Three fun facts about their organisms (Byrne 2013)
5. At least two questions they have about their organisms.

Instructors must also print one sign per soil organism group to identify stations set up around the classroom where students will present their research findings (one station per assigned organism; Box 11.1). To expand and enrich the activity, live or preserved specimens corresponding to the assigned organisms can be included at these stations. To prepare, the instructor can purchase (e.g., from a biological specimen distributor), collect from the field, or culture (e.g., bacteria and fungi) live soil organisms to populate stations. Preserved specimens (e.g., springtails and mites) may also be used and can be also purchased. Alternatively or in addition, if the class has access to outdoor space with soil, students can participate in scavenger hunts in pairs prior to this lesson to find soil organisms, identify and research them as described above, and bring them to class for the activity (Nardi 2007 is an excellent resource for identifying soil organisms in the field). As appropriate and possible, stations should also be equipped with proper instrumentation for students to observe the organisms (e.g., dissecting and compound microscopes, microscope slides, dissecting probes, Petri dishes, etc.) If materials to observe microscopic organisms are not available, instructors can choose and assign solely macroscopic organisms.

A chalk- or whiteboard or flipchart paper will be needed for constructing the soil food web diagrams following the station activity. Instructors should also prepare and print educational level-appropriate worksheets for each student to complete during the activity; suggested high school, college, and graduate school worksheets are provided in ESM-B, ESM-C, and ESM-D.

This activity may be modified for larger or smaller classes by adjusting the number of organisms or for different age levels by modifying the sophistication with which the organisms, their relationships, and their functions are addressed and discussed (as in ESM-B, ESM-C, and ESM-D). For example, for smaller classes, fewer organisms can be used while keeping the structure of the activity the same. Alternatively, students can each become an "expert" on two organisms, and the activity can be run in four rounds, with each student acting as an "expert" for two rounds and a "learner" for two rounds (see activities' description below). For larger classes, more organisms can be added (e.g., lower-level taxonomic groups), and/or the activity can be run as two or more simultaneous rotations, wherein each rotation occurs in a separate space in the classroom and has two "experts" for each organism. To increase the sophistication of the activity, upper-level undergraduates and graduate students may also be required to find, read, and bring to class one peer-reviewed journal article pertaining to their organism. Instructors may then require these students to provide an annotated summary of the paper, summarize the paper for their classmates, or reflect and report upon how human land use decisions, climate change, or some other disturbance may affect their soil organisms and vice versa.

Activities

On the day of the activity, the instructor sets up stations throughout the classroom, each designated by a sign denoting the assigned organisms. The instructor should initiate the activity with a brief introduction to soil biodiversity (see ESM-A for suggested resources) and, where appropriate, refer to prior lessons about biodiversity, food webs, energy exchange, and the role of organisms in ecosystem processes and services. Using their "expert" information cards prepared before class, students are invited to engage in 20-min rounds of discovery, in which each student has the opportunity to be the representative "expert" teacher for their prepared organism. Students will spend the other round(s) as information-seeking learners, when they will be given 20 min to visit each station. Students are instructed to go to their assigned stations and decide which of the student pair will first be the "expert" and which will be the learner. When the instructor indicates the start of the activity, "experts" remain at their stations as the learners informally travel around the room from station to station. As the learners approach each station, "experts" teach about their organism using the information on their cards, images, and/or videos and learners are encouraged to ask questions while completing the assigned worksheets (ESM-B, ESM-C, and ESM-D). After 20 min, the instructor signals students to switch, at which point the "experts" become the learners and vice versa, and the new group of learners circulates among the different stations.

After visiting the organism stations, the instructor guides students in assembling their cards into a soil food web diagram by taping cards to a large piece of paper or board and drawing arrows among organism groups to denote feeding relationships and energy transfer relationships (see Byrne 2013 for details, including an example web diagram). The activity ends with a 10–15 min wrap-up discussion in which the instructor has students share "fun facts" they found fascinating or "mind-blowing," in addition to questions the activity raised. The instructor can document student questions and observations on the board for follow-up inquiry and discussion. In this share-out, the instructor should help students deepen and apply their learning by helping them to identify the relationships among soil organisms and between soil organisms and other ecology and sustainability topics such as ecosystem functioning, provisioning of ecosystem services, agricultural productivity and sustainability, food security, plant ecology, climate change, and human well-being (see Box 11.2 for suggested questions). Instructors may wish to choose only one or two

---

Box 11.2. Wrapping Up: Questions for Discussion

- Which organisms you observed depend upon one another and how?
- What would happen to the food web if one soil organism or functional group were removed or added?
- How do soil organisms affect ecosystem functioning, and what ecosystem services do soil organisms provide?
- How might certain land management decisions (e.g., deforestation, tillage, development, restoration) affect the soil food web?
- How could alteration of the soil food web affect agricultural productivity and sustainability?
- How do soil-mediated ecosystem functioning and ecosystem services affect human societies and well-being?
- How might changes in aboveground dynamics (e.g., deforestation, plant invasion) affect the soil food web?
- What are some feedbacks between soil organisms and regional and global climate?

applied topics to focus on for discussion to make the activity as applicable as possible to their course content and curriculum and to honor student level and sophistication (see ESM-A for suggestions).

Follow-Up Engagement

- Students can compare soil organisms and food webs across various biomes (e.g., tundra, taiga, tropical, etc.) by researching a biome-specific soil organism, creating biome-specific food webs, and then discussing their similarities and differences and their respective effects on ecosystem processes and services.
- Students can visit a local farm to talk with farmers about how they manage the soil. Beforehand, students can construct an agricultural soil food web and discuss the effects of different farm management strategies on soil organisms and vice versa. Students can be asked to read and summarize articles about soil organisms and sustainable agricultural productivity and then prepare questions for the farmer about agricultural land management.
- Instructors can plan a field trip to a local forest for a follow-up soil scavenger hunt (ESM-E, ESM-F) that will give students an opportunity to find new soil organisms and engage in extended discussions about relationships among plants, soil organisms, other ecological variables, and sustainability (see Box 11.2 for suggested discussion points; also see Chap. 10, this volume, for related activities).
- Instructors may assign additional readings to stimulate and guide discussions about relationships between soil organisms and global climate change, global nutrient cycling, aboveground-belowground dynamics, agricultural sustainability, and socioeconomic factors (see ESM-A for suggested resources).
- Students may conduct and summarize a literature review about their organisms in a short blog entry, some other form of social media, or a brief research paper that makes connections between their organism(s) and a chosen applied topic (see Box 11.2 for suggestions).

Connections

- Soil organisms play a central role in ecosystem ecology and global biogeochemistry and thus profoundly influence nutrient cycling, primary productivity, and climate change. Understanding the complexity of belowground food webs and potential implications of human impact upon them enriches students' ecological values, encouraging them to incorporate a broader worldview in approaching issues around climate change.
- Soil organisms have critical roles in the sustainability of natural and agricultural systems, food systems, human well-being, and community resilience. As such, a disruption in the balance of food webs has implications that extend to political, economic, and social concerns.

- Land use decisions (e.g., agriculture, deforestation, development, restoration) have complex, site-specific effects on soil food webs and their effects on ecosystem sustainability and vice versa. Understanding the complex relationships of belowground organisms can help deepen the attitudes and values of students to encompass a more holistic worldview in assessing the impacts of land use decisions (e.g., see Chaps. 12 and 20, this volume).

Acknowledgments Thank you to all of the Antioch University New England Soil Ecology students for providing feedback that has informed and inspired these activities.

References

Bardgett RD, Wardle DA (2010) Aboveground-belowground linkages: biotic interactions, ecosystem processes, and global change. Oxford University Press, Oxford

Byrne LB (2013) An in-class role-playing activity to foster discussion and deeper understanding of biodiversity and ecological webs. EcoEd Digital Library, http://ecoed.esa.org

Nardi JB (2007) Life in the soil: a guide for naturalists and gardeners. The University of Chicago Press, Chicago and London

Schimel J (2013) Soil carbon: microbes and global carbon. Nat Clim Change 3(10):867–868

Wagg C, Bender SF, Widmer F, van der Heijden MGA (2014) Soil biodiversity and soil community composition determine ecosystem multifunctionality. Proc Natl Acad Sci U S A 111:5266–5270

Wall DH, Six J (2015) Give soils their due. Science 347:694

Chapter 12
Fire, Pollution, and Grazing: Oh My! A Game in Which Native and Invasive Plants Compete Under Multiple Disturbance Regimes

Heather E. Schneider and Lynn C. Sweet

Introduction

Human activities have facilitated the spread of plant propagules throughout the world, leading to the introduction of exotic species to new environments. Some of these exotics will become invasive and alter ecosystems in numerous, sometimes negative, ways. Invasive plants can significantly degrade wildlife habitat, alter soil nutrient cycling, increase the risk of wildfires and floods, consume large amounts of water, and disrupt other ecosystem services (Vilà et al. 2011; also see Chaps. 9–11, this volume). Such problems may require costly mitigation or management by humans (e.g., managing invasive plants in grazing and agricultural systems in the USA costs $30 billion per year) (Hejda et al. 2009, Cal-IPC 2014, van Andel and Aronson 2012). In some instances, managing invasives may be difficult or impossible depending on the scale of the invasion. Furthermore, environmental conditions created by disturbances (e.g., flood, fire) and anthropogenic impacts (e.g., pollution, livestock grazing) can alter the competitive balance between native and invasive plants and complicate management efforts (see additional references in Chap. 9, this volume).

The activity described here is an interactive game designed to introduce students to the causes, consequences, and management of exotic plant invasions in the context of different ecosystem perturbations. By playing the game, the goal is for

Electronic supplementary materials: The online version of this chapter (doi:10.1007/978-3-319-28543-6_12) contains supplementary material, which is available to authorized users.

H.E. Schneider (✉)
Department of Ecology, Evolution, & Marine Biology, University of California, Santa Barbara, CA, USA
e-mail: heather.schneider@lifesci.ucsb.edu

L.C. Sweet
Center for Conservation Biology, University of California, Riverside, CA, USA

© Springer International Publishing Switzerland 2016
L.B. Byrne (ed.), *Learner-Centered Teaching Activities for Environmental and Sustainability Studies*, DOI 10.1007/978-3-319-28543-6_12

students to learn about plant community dynamics with a focus on competition between invasive and native plants, the impacts of disturbances on interspecific interactions, and the challenges and threats facing the management and conservation of native vegetation and ecosystem services. Although the game's materials were designed for ecosystems characteristic of Southern and Central California (presented here as a case study), the materials can be adapted for any regional flora.

Learning Outcomes

After completing this activity, students should be able to:

- Define and differentiate among the characteristics of native, exotic, and invasive species.
- Define and distinguish among the effects of anthropogenic impacts and natural disturbances in ecosystems (e.g., fire disturbance, pollution, grazing).
- Explain how invasive plants can alter ecosystem structure and function.
- Analyze how different environmental conditions created by disturbances and human activities might change the local abundance and spatial extent of native and invasive plants.
- Design a management scenario that conserves native species and ecosystem services by reducing invasive species abundance.

Course Context

- Developed for 20–25 sophomore high school students in a survey course covering a variety of topics in basic botany and plant ecology. Best suited for classes of 10–50 students
- 60 min in one class meeting
- Before class, students read the case study and fill out a vocabulary worksheet. In class, the suggested introductory presentation provides the necessary background for the activity
- This activity could be adapted for a college course by providing more in-depth information about the species and increasing the difficulty of challenge questions, as well as assigning a research project outside of the classroom (included in the "Follow-Up Engagement" section)

Instructor Preparation and Materials

Instructors should familiarize themselves with basic ecological concepts involving ecosystem structure and function (e.g., Chaps. 14–15 in Gurevitch et al. 2006; Section 3.4 in Sekercioglu 2010), natural disturbance of and anthropogenic impacts on native ecosystems, invasive plants (Mack et al. 2000; Cal-IPC 2014), and invasive plant management (van Andel and Aronson 2012; Cal-IPC 2014; also see Electronnic Supplementary Materials (ESM) A for additional resources and basic content relevant to the lesson). The instructor may also wish to research local native and invasive species using online references (see ESM-A) to adapt the activity to be regionally specific. The instruction guide (ESM-B) provides step-by-step instructions for implementing the activity and notes for modifications. Instructors should review and, as needed, modify the provided presentation slides (see Electronic Supplementary Materials on website; ESM-A) and vocabulary/case study background worksheet (ESM-C). Students should be given the vocabulary and case study background handout (ESM-C, printed out double sided) at a class meeting prior to the activity and asked to research and fill out the definitions before class so that they are ready to share them with their peers during the introductory presentation.

A set of activity cards for the game (ESM-B includes examples from a Southern California ecosystem; see ESM-B for more information on how to play the game) should be printed double sided, in color on 8.5×11″ paper, one card per student, plus a few extras; for durability, they can be printed on cardstock or laminated. Instructors may modify the species and disturbances on the cards in ESM-B to match local ecosystems, if desired.

This activity requires open floor space so that students can segregate into groups and move about the classroom. If space is limited, students can participate from their desks and stand or sit, depending on whether or not they are actively participating in the game. Similarly, students could line up around the perimeter of the room if open floor space is not available.

Activity

First, the instructor goes through the introductory presentation (see the introductory slides 2–18 in ESM-A), introducing the concepts of invasive and native species and disturbances (~20 min), and provides an overview of the game to the students (slides 19–23), before beginning the game. The game can be summarized as an activity that will interactively demonstrate the fluctuations in native and invasive plant abundance in a community that is experiencing natural disturbances or anthropogenic impacts. The goal of the activity is for students to visualize changes in communities of plants containing invasive and native species, while learning how populations may change based on disturbances and anthropogenic impacts. The students will act *as* plant species for the activity, responding to environmental changes based on information they are given. (For this point, it may be helpful, and fun, to encourage students to "get into character" by imagining themselves as their assigned plant.) The events in the game mirror those that affect the dynamics of real plant populations and communities: survival, mortality, competition, and reproduction.

Next, the students will play the game, in two rounds, with each round taking 5–10 min, depending on the class size. (Additional rounds, which may have different results and enhance student learning, may be played if time allows.) Each round consists of all students moving around the classroom, recording population data (by filling out the charts starting on slide 24), and answering challenge questions (from ESM-B) based on introductory material. Briefly, the game proceeds as follows (see ESM-B for detailed instructions):

1. After introducing the game (see above and slides 19–23), students are assigned the identity of either a native or invasive plant species based on the description on the card they are given (ESM-B). These will form the basis of creating two teams: Natives and Invasives.
2. For Round 1, go through steps (a) and (b) (below) for the natural disturbance regime. Then for Round 2, repeat these steps with the anthropogenic regime (ESM-B contains a script for the instructor to guide these steps):

   (a) The instructor will read the name of a disturbance or population-level event (competition, reproduction), resulting in students either remaining standing (survival), sitting down (mortality), or rejoining the group (reproduction), depending on the biological response to the event stated on their card.
   (b) As the game progresses, the instructor should tally the number of native and invasive individuals in play on the table provided (slides 24–25) or into the fillable charts on the next slide (slides 26–27). Students should be asked to take note of the changes in the total population of natives vs the total population of invasives and discuss them after the game.

3. After two rounds (or more if time allows), the wrap-up "Questions for Discussion" in the presentation (slide 28) can be used to recap the activity with the whole class. Students may answer questions as a group or through a "think-pair-share" exercise.

Follow-Up Engagement

- The following prompts can be used for extended discussions, take-home assignments, and assessment of learning:
  - Can the outcome of individual plant competition affect the ultimate species composition of a community? If so, in what circumstances are these competitive interactions most important?
  - Are some species better able to quickly "take over" than others? Which biological event or species characteristic (reproduction, competitive success, or disturbance survival) seems to have made the most difference in this activity?
  - List the steps of a species' invasion (from slide 10 in ESM-A).
  - What is the difference between an anthropogenic disturbance and a natural disturbance? How can anthropogenic impacts affect natural disturbances?

- Give an example of two species from the game that responded differently to the same disturbance. Why might they have responded differently?
- Draw a graph depicting the "Intermediate Disturbance Hypothesis" (from slide 15 in ESM-A) and explain the theory behind it.
- Describe two examples of anthropogenic and natural disturbances. Research two examples relevant to an ecosystem in the state or region you live in and summarize how each example impacts plant species.
- Design a scenario that combines natural disturbances and anthropogenic impacts with direct management techniques to promote native species success and reduce invasive species abundance. Research the invasive species and cite sources supporting your plan for disturbance and management and why this plan is predicted to work. Also, describe the plan's expected costs, required labor, and timeline (e.g., will invasives be controlled after one treatment or will this be a multiyear effort?).

Connections (See ESM-A for References About These Topics)

The issue of invasive species can be discussed in the context of other environmental and sustainability topics, including:

- Land-use change, such as urban development, can cause ecosystem changes that favor invasive species over natives. Roads can create corridors of habitat connectivity that facilitate the spread of invasive species from one area to another.
- Air, water, and soil pollution can affect species invasions. For example, anthropogenic nitrogen pollution, from the burning of fossil fuels and fertilizer applications, alters nutrient availability, which may be disproportionally beneficial to invasive species (also see Chap. 9 for references).
- Climate change can enhance the problems caused by invasive species. For example, in arid and semiarid ecosystems, the increased frequency and intensity of wildfires due to climate change can increase invasive plant abundance.
- Management of invasive species and restoration of natural systems are also a social and moral issue. Both activities have major economic impacts and inspire philosophical debates about whether invasives should be removed or accepted as part of a changing ecosystem.

Acknowledgments The authors would like to thank the Center for Science and Engineering Partnerships at the University of California, Santa Barbara for funding the development of this activity as part of the School for Scientific Thought program. LC Sweet was supported in part during the preparation of this contribution by the National Science Foundation, NSF #EF-1065864, and HE Schneider was supported in part by NSF #DEB-1142784.

References

Cal-IPC (2014) The impact of invasive plants. California Invasive Plant Council. http://www.cal-ipc.org/ip/definitions/impact.php. Accessed 11 Dec 2014

Gurevitch J, Schneider SM, Fox GA (2006) The ecology of plants, 2nd edn. Sinauer Associates, Inc., Sunderland, MA

Hejda MP, Pysek P, Jarosik V (2009) Impact of invasive plants on the species richness, diversity and composition of invaded communities. J Ecol 97:393–403

Mack RN, Simberloff D et al (2000) Biotic invasions: causes, epidemiology, global consequences and control. Issues Ecol 5:1–20

Van Andel J, Aronson J (2012) Restoration ecology: the new frontier, 2nd edn. Wiley-Blackwell, Hoboken, NJ

Vilà M, Espinar JL, Hejda M et al (2011) Ecological impacts of invasive alien plants: a meta-analysis of their effects on species, communities and ecosystems. Ecol Lett 14:702–708

Sekercioglu CH (2010) Ecosystems functions and services. In: Sodhi NS, Ehrlich PR (eds) Conservation biology for all. Oxford University Press, London (Available online at: https://conbio.org/publications/free-textbook)

Chapter 13
Exploring Trophic Cascades in Lake Food Webs with a Spreadsheet Model

Kyle A. Emery, Jessica A. Gephart, Grace M. Wilkinson, Alice F. Besterman, and Michael L. Pace

Introduction

An ecosystem is a space in the environment, of any size, in which the living and nonliving components of that environment interact. The living organisms form food webs, which describe the transfer of energy and nutrients through trophic linkages (e.g., predation). The stability of an ecosystem's food web, which is necessary for conserving its species or feeding human populations, depends on both biological and physical factors (see Chaps. 10 and 14, this volume, for more background). Food webs are often at a greater risk of change when affected by anthropogenic disturbances, such as reduced biodiversity (Pace et al. 1999). Lakes, for example, are important sources of freshwater, food, and recreation while being highly dependent on inputs and management-related decision making. A change at one trophic level of a lake food web can result in a trophic cascade, during which the population size of other levels is altered (Carpenter et al. 1985; Pace et al. 1999; Polis et al. 2000). These changes can arise when changes in higher levels then affect lower ones (a top-down cascade) or vice versa (a bottom-up cascade). The cascading effects can have consequences for both the system itself and the humans that depend on it with consequences for ecosystem management, ecosystem services, and biodiversity conservation.

Understanding trophic cascades is important for students who are learning about how people's decisions affect the environments in which they live. The goal of this learning activity is for students to discover how anthropogenic perturbations can induce trophic cascades in a lake food web. Students will

Electronic supplementary materials: The online version of this chapter (doi:10.1007/978-3-319-28543-6_13) contains supplementary material, which is available to authorized users.

K.A. Emery (✉) • J.A. Gephart • G.M. Wilkinson • A.F. Besterman • M.L. Pace
Department of Environmental Sciences, University of Virginia, Charlottesville, VA, USA
e-mail: Kae2n@virginia.edu

explore multiple scenarios using a food web spreadsheet model and gain experience in interpreting model output. This activity will also engage students in hypothesis-driven thinking and subsequent problem solving through synthesis questions focused on responsible decision making about sustainable ecosystem management.

Learning Outcomes

After completing this activity, students should be able to:

- Describe a lake food web and trophic cascades within it.
- Distinguish between top-down and bottom-up controls on food webs.
- State hypotheses for and interpret output from food web model scenarios.
- Evaluate strategies to prevent and manage the effects of trophic cascades on biodiversity and ecosystem services.
- Predict causes of trophic cascades in diverse ecosystems and their associated consequences for ecosystem services and human health.

Course Context

- Developed for an intro or advanced ecology or biology course with any number of students
- 50 minutes in one class meeting
- No background knowledge is needed; instructor presents necessary information prior to activity

Instructor Preparation and Materials

Using this activity requires general knowledge of ecosystem structure and function, specifically food webs and trophic linkages. The sources cited in the introduction will prepare instructors for this specific activity, but additional general background can be found in Weathers et al. (2012) and Terborgh and Estes (2010); for more on trophic cascades specifically, see Estes et al. (2011), Ripple and Beschta (2012), and Bunnell et al. (2014). Instructors can use the instructor guide, student worksheet, and glossary (see electronic supplementary materials (ESM) A, B, and C) to orient themselves to the model (ESM-D) and in turn help students learn how to use it and obtain their results. Worksheets and the glossary should be modified as needed (e.g., to match student background, to remove synthesis questions for later exam) and printed for each student or group of students. Instructors should plan for how students will interact with the model and how it will be disbursed (by email or flash drive). Ideally groups of students will work on their own computers to allow for independent hands-on student-driven hypothesis testing (if not, instructors may walk students through the model as it is projected to the class).

The instructor may use or modify the slides (ESM-E) to present background information on the science (concepts, definitions) and the model (examples, scenarios, and associated questions). The background material in these slides will provide students with an overview of the necessary concepts for this activity, but the instructor may revise these depending on the students' background and degree to which the extension activities and synthesis are utilized. See the supplementary instructor guide (ESM-A) for more details about using and editing the slides.

Activities

To prepare students for working with the food web model, the instructor should begin with a presentation (10–15 min) of relevant background material including how to manipulate the model (see ESM-A). Then, students, individually or in groups, will run simulations with the food web model (ESM-D) through four scenarios (5–7 min each) that can cause trophic cascades through either top-down or bottom-up control: overfishing, stocking fish, fertilizer runoff, and invasive species. Students will read through each scenario description and complete the worksheet (ESM-B) as they proceed (hypotheses and observations can also be recorded in the model file). As the students work, the instructor can help students navigate the model and interpret the figures (see guidance in ESM-A).

For each scenario, students should read the description located in the model and on the student worksheet. Students should then write down a hypothesis stating their prediction of what will happen to the food web given the described manipulation in the scenario. As desired, the instructor can initiate a brief whole-class discussion regarding the students' predictions before they use the model. After making their predictions, students should manipulate the model using low, medium, and high values from the range of possible values supplied on the spreadsheet. The instructor should walk around the class to engage with the students, discussing the simulations and addressing any questions. Students should interpret the graphical model output and record their observations on the worksheet. After students have made their observations, they should answer the synthesis questions for that scenario. Responses to these questions may be discussed as a group before moving on to the next scenario. These steps should then be repeated for the subsequent three scenarios.

As the class progresses through each scenario, the instructor should reemphasize the central ecological concepts (ESM-C and E). Questions can be posed to ask for explanations of how key concepts such as eutrophication or top-down control relate to the scenarios. The instructor guide (ESM-A) provides additional guidance and examples. The instructor can also extend classroom discussion with the synthesis questions for each scenario, or other situations or ecosystem types (see the Follow-Up Engagement section). To conclude the activity, the instructor should facilitate a discussion (5–10 min) about what the students learned and how understanding human-caused trophic cascades can contribute to responsible ecosystem management. Students should have gained an understanding of how ecosystems can be altered in many ways and how trophic cascades can shift food webs to new compositions, with consequences for both the ecosystem and those who depend on its services.

Follow-Up Engagement

The following prompts could be used for extended discussion, follow-up assignments, or assessment (e.g., reflection papers or exam questions); also see the synthesis and extension questions in ESM-B:

- What are some features of or benefits from the lake that are not in this model that a manager should consider?
- If you managed this lake or its watershed, how would you address the potential drivers of a trophic cascade (e.g., point vs. nonpoint source pollution controls, regulate recreational activities, etc.)? Summarize a lake management plan that balances the considerations of each scenario. Justify your management decisions. (For a related activity, see Chaps. 16, 19, and 20, this volume.)
- What roles do ecology and ecological modeling play in management? What or who else influences management decisions? How should we balance scientific information with other considerations such as diverse stakeholder interests?
- Find a real world example of each scenario explored in class. What caused the trophic cascade? How was the situation remedied, if at all? (See ESM-A for examples.)
- Choose a different ecosystem and develop a simple food chain for that system (with four or more species). Identify potential anthropogenic alterations to the system that would lead to a trophic cascade. Hypothesize how these activities will affect the ecosystem and how the modified food web will appear.

Connections

- This activity highlights possible trade-offs between economic and environmental gains in decision making for the various stakeholders involved.
- Ecosystem concepts are introduced that can be further developed in future lessons. Refer to Glossary concepts (ESM-C) when appropriate to reinforce the ecosystem-level ideas covered in this activity. (Also see Chaps. 10, 12, and 14–16 for related activities.)
- In the context of land- and water-use lessons, this activity helps demonstrate how human management of the environment can be used to promote sustainable eco-

system management or negatively impact biodiversity, system functionality, and ecosystem services in the context of food production, the effects of invasive species, and pollution (e.g., see Chaps. 12, 14, and 15–17, this volume).

- This activity can be referenced as an example of evidence-based decision making. Using data to guide decision making is relevant in both science and policy fields (e.g., see Chap. 34, this volume).

Acknowledgments This contribution was supported by NSF grants DEB-1456151 and DEB-1144624, the NSF GRFP, and the University of Virginia Department of Environmental Science.

References

Bunnell DB, Barbiero RP, Ludsin SA et al (2014) Changing ecosystem dynamics in the Laurentian Great Lakes: bottom-up and top-down regulation. Bioscience 64:26–39

Carpenter SR, Kitchell JF, Hodgson JR (1985) Cascading trophic interactions and lake productivity. Bioscience 35:634–639

Estes JA, Terborgh J, Brashares JS et al (2011) Trophic downgrading of planet earth. Science 333:301–306

Pace ML, Cole JJ, Carpenter SR, Kitchell JF (1999) Trophic cascades revealed in diverse ecosystems. Trends Ecol Evol 14:483–488

Polis GA, Sears ALW, Huxel GR, Strong DR, Maron J (2000) When is a trophic cascade a trophic cascade? Trends Ecol Evol 15:473–475

Ripple WJ, Beschta RL (2012) Trophic cascades in Yellowstone: the first 15 years after wolf reintroduction. Biol Conserv 145:205–213

Terborgh J, Estes JA (eds) (2010) Trophic cascades: predators, prey, and the changing dynamics of nature. Island Press, Washington, DC

Weathers KC, Strayer DL, Likens GE (eds) (2012) Fundamentals of ecosystem science. Elsevier, Waltham, Massachusetts

Chapter 14
Ants, Elephants, and Experimental Design: Understanding Science and Examining Connections Between Species Interactions and Ecosystem Processes

Christine M. Stracey

Introduction

Students often view the information in textbooks as static facts they must memorize. In contrast, scientific knowledge is extremely dynamic because the information we have today can be rapidly refined and even replaced as we learn more about the world around us, while much of our knowledge of any given topic is inherently incomplete and therefore uncertain. Without an understanding of the dynamic process of science and how science is conducted, students can become frustrated with the seemingly contradictory information in the news and be misled when scientific issues are debated and become politicized. This is further complicated by the fact that decisions are often made based on incomplete information from preliminary studies and may in hindsight be judged as "wrong" when new information is acquired. The dynamic nature of science is particularly important to understand when science is used to make decisions about complex issues involving the environment and sustainability. Therefore, it is important for students learning about these topics to understand the scientific process and be able to effectively critique different experimental designs, understand the strengths and limitations of any given study, and recognize the roles that incomplete information and uncertainty play in science (also see Chaps. 9 and 24, this volume).

Electronic supplementary materials: The online version of this chapter (doi:10.1007/978-3-319-28543-6_14) contains supplementary material, which is available to authorized users.

C.M. Stracey (✉)
Department of Biology, Guilford College, Greensboro, NC, USA
e-mail: straceycm@guilford.edu

© Springer International Publishing Switzerland 2016
L.B. Byrne (ed.), *Learner-Centered Teaching Activities for Environmental and Sustainability Studies*, DOI 10.1007/978-3-319-28543-6_14

To address student understanding of the nature of science, an activity was developed using a video abstract from a paper (Goheen and Palmer 2010) that examines the role of mutualisms in shaping ecosystem properties. Positive species interactions (e.g., mutualisms) have long been treated as less important than negative species interactions (e.g., competition) in structuring ecological communities, but recently have been found to have far-reaching consequences for both communities and ecosystems (Stachowicz 2001; Bruno et al. 2003; Kiers et al. 2010). Species interactions are often taught separately from ecosystem ecology (e.g., nutrient cycling), despite the interrelated nature of all ecology. The goals of this learning activity are for students to (1) evaluate and critique experimental design, (2) understand how interspecific interactions can scale up to have ecosystem level effects, and (3) for students to recognize and relate to the personal aspect of science by watching a scientist describe his study.

Learning Outcomes

After completing this activity, students should be able to:

- Interpret data from published graphs.
- Identify key components of experimental design.
- Discuss inferences that can be made from results generated by different experimental designs.
- Identify and describe interspecific interactions.
- Explain how interactions at one level of ecological organization (e.g., communities) can influence processes at other levels (e.g., ecosystems).

Course Context

- Developed for in an introductory biology course for majors with 15–30 students for use in a unit on ecosystems that followed one on community ecology
- 110 min in one class period. The inclusion of introductory and discussion material would take additional time or adjustments can be made for a shorter implementation
- To enhance their success in this activity, students should be familiar with interspecific interactions and ecosystem processes of nutrient cycling. In addition, students are likely to struggle with the activity if they have not been exposed to basic concepts of experimental design, which was a recurring topic throughout the semester in the course for which this activity was originally designed. Instructors could assign students Chap. 1 from Ruxton and Colegrave (2011) as background reading as appropriate. Students struggle the most with replication and being able to provide numerous examples helps students. A familiarity with

the strengths and weaknesses of both correlational and manipulative studies prior to this activity (addressed in Chap. 2 of Ruxton and Colegrave 2011) helps students think about the benefits of multipart experiments

Instructor Preparation and Materials

The instructor should be familiar with species interactions, ecosystem processes, and the basic principles of experimental design. Introductory biology textbooks provide sufficient background on community and ecosystem ecology. Brief discussions of positive interactions can be found in Stachowicz (2001) and Bruno et al. (2003), while an introduction to ecosystem ecology can be found in Chapin et al. (2002). (Also see Chaps. 10, 12 and 13 for more about ecosystems and species interactions.) For more background on experimental design, see Electronic Supplementary Materials (ESM-A), Ruxton and Colegrave (2011), and Slutz and Hess (2015). For a good discussion of issues of replication in ecological studies, see Hurlbert (1984).

Prior to the activity, instructors should download the other ESM and familiarize themselves with their content, editing them as needed to match their course needs. The presentation file (ESM-B) does not contain pictures due to copyright issues, so it is suggested that instructors add pictures to make the presentation more visually appealing. Instructors should watch and download the video (link provided on slide 3 in ESM-B) and read the primary paper that it is based on (Goheen and Palmer 2010) to prepare for answering students' questions and extending the discussion as needed. ESM-C contains an instructor script that outlines the times in the video to pause along with accompanying discussion questions and suggested answers. ESM-D contains a student worksheet with the discussion questions that should be printed for students to complete in class. Instructors should help students prepare for this activity by ensuring they have sufficient background understanding of key concepts (e.g., through lecture, homework, or other lessons).

Activities

As context for the exercise, the instructor could review Table 1 with students at the start of the activity by providing it as a handout or projecting it on a board such that students can reference it throughout the activity. For the activity, students watch a 5 min video abstract from a recently published paper and work in small groups during pauses in the video to answer questions about the study. Following this small group work, the class discusses these questions and then proceeds to watch the next video segment. Before showing the video, instructors could give a brief introduction to the activity (e.g., challenge them to think about how interspecific interactions can affect ecosystems, review experimental design using ESM-A, etc.). Alternatively, the video can be shown without prelude as an attention-grabbing device. Hand out the student worksheet (ESM-D) and show the video while pausing at seven points as indicated in the instructor script (ESM-C). At each pause, display the corresponding presentation slide (ESM-B) and have students work together in groups of two to four to answer the respective questions on their worksheet. Give students time (3–6 min) to discuss each set of questions in their groups while circulating and providing assistance as needed. It is important to allow students sufficient time to struggle with the experimental design and graph questions before coming back as a class and discussing them. Students will usually need more time for question sets about experimental design and less time for introductory questions. Question sets that require longer time for student work are indicated in the instructor script. The worksheet has two sections for each question: one for their answers that they derive from their small group discussions and another for any revisions to their answers after the class discussion. Encourage students not to change their initial answers, but use them as a learning tool that helps them reflect on and clarify their thinking. The summary questions on the worksheet (and on presentation slide 16) can be used to conclude the activity, or if time permits, suggestions from the Follow-Up Engagement and Connections sections (below) can be used to extend the activity. The student worksheet can also be collected and used to assess the learning outcomes.

For shorter in-class implementation, students could be given the graphs—or for more advanced students, the original paper—to review as homework before class during which they write open-ended summaries for each graph or identify the dependent and independent variables and the hypotheses being tested. Alternatively, the questions could be used as "clicker questions" embedded in the presentation slides that students respond to with handheld devices, which could also be used as a modification for larger class sizes.

Follow-Up Engagement

- Ask students to think about why most conservation is focused on so-called charismatic megafauna, like elephants, while very little attention is given to invertebrates, such as ants. What are some possible arguments for focusing on charismatic megafauna? Using the results of this study, ask them to make an argument for paying more attention to less charismatic species. (For a related activity, see Chap. 18, this volume.)
- Students can be assigned an activity in which they propose a follow-up experiment addressing a conservation issue that includes all the elements of experimental design (see ESM-A). Students can then peer review these, and the class can discuss the strengths and limitations of different designs.
- Have students critique the news story regarding this study (Bhanoo 2010). Did it leave out important information? Does it help explain aspects of the study that were unclear from the video abstract? (See Chap. 37, this volume, for a relevant activity.)

- Have students read the newspaper article (Dean 2008) about a different study of the effects of ant-plant mutualisms (Palmer et al. 2008) and then respond to the following prompts:

  - Describe multiple effects of the loss of large herbivores on an African savanna.
  - Using the graphs from the original paper, support the claims made in the newspaper article.

Connections

- Anytime an experimental study is discussed, students can use the experimental design framework (ESM-A) to evaluate it, from identifying its design components to critiquing the validity of the design and the reliability of the results and conclusions drawn.
- Ecological concepts, such as biotic and abiotic habitat requirements, limiting factors, niches, and biogeography, all connect with the variation in distribution of ant-acacia plants and elephants examined in Goheen and Palmer (2010). Further, this study can be linked to discussions about nutrient cycling, fire regulation, other ecosystem services, and conservation that can be affected by the presence of different organisms and their interactions in an ecosystem (e.g., Kiers et al. 2010; also see Chaps. 10 and 12, this volume).
- This activity provides an example of the importance of ecological research for making management and policy decisions. The example study demonstrated that the management of elephant populations and the use of exclosures can have unintended consequences on ecosystem composition and function.

References

Bhanoo S (2010) In a fight for a tree, ants thwart elephants. New York Times. http://www.nytimes. com/2010/09/07/science/07obants.html?_r=0. Accessed 31 Dec 2014

Bruno J, Stachowicz J, Bertness M (2003) Inclusion of facilitation into ecological theory. Trends Ecol Evol 18:119–125

Chapin FS, Matson P, Mooney H (2002) Principles of terrestrial ecosystem ecology. Springer-Verlag New York, Inc., New York, NY. Available online http://www.crc.uqam.ca/Publication/ Principles%20of%20terrestrial%20ecosystem%20ecology.pdf

Dean C (2008) In life's web, aiding trees can kill them. New York Times. http://www.nytimes. com/2008/01/11/science/11ants.html. Accessed 31 Dec 2014

Goheen J, Palmer T (2010) Defensive plant-ants stabilize megaherbivore-driven landscape change in an African savanna. Curr Biol 20:1768–1772

Hurlbert S (1984) Pseudoreplication and the design of ecological field studies. Ecol Monogr 54:187–211

Kiers ET, Palmer T, Ives A et al (2010) Mutualisms in a changing world: an evolutionary perspective. Ecol Lett 13:1459–1474

Palmer T, Stanton M, Young T et al (2008) Breakdown of an ant-plant mutualism follows the loss of large herbivores from an African savanna. Science 319:192–195

Ruxton G, Colgrave N (2011) Experimental design for the life sciences, 3rd edn. Oxford University Press, Oxford

Slutz S, Hess KL (2002–2015) Experimental design for advanced science projects. http://www.sciencebuddies.org/science-fair-projects/top_research-project_experimental-design.shtml. Accessed 21 Aug 2015

Stachowicz J (2001) Mutualism, facilitation, and the structure of ecological communities. Bioscience 51:235–246

Chapter 15
Teaching Lyme Disease Ecology Through a Primary Literature Jigsaw Activity

Jae R. Pasari

Introduction

Lyme disease, a bacterial infection transmitted via ticks, is the most common and fastest spreading vector-borne disease in the United States (Bacon et al. 2008). Since the early 2000s, research has increased on the reasons causing its range expansion. However, a scientific consensus about the mechanisms leading to its spread has been slow to emerge. Instead, many plausible hypotheses have been proposed, tested, and debated. Experiments have helped to uncover the roles of forest fragmentation and regrowth, host community diversity, the loss of apex predators, invasive species, episodic oak masting, and increasing deer populations (Ostfeld 2010, 2013).

The activity in this chapter offers students the opportunity to assess the Lyme disease debate with primary literature using a jigsaw group discussion format. The goals of this activity are to increase students' abilities to (1) read and comprehend primary scientific literature; (2) summarize, explain, and synthesize complicated data; and (3) identify gaps in contemporary research to form plausible hypotheses that can guide future research. In addition, students should deepen their knowledge of ecological concepts related to populations, predator–prey dynamics, trophic cascades, parasite ecology, and competition (see Chaps. 12–14, this volume, for related topics and activities).

Electronic supplementary materials: The online version of this chapter (doi:10.1007/978-3-319-28543-6_15) contains supplementary material, which is available to authorized users.

J.R. Pasari (✉)
Science Department, Berkeley City College, Berkeley, CA, USA

Science Department, Pacific Collegiate School, Santa Cruz, CA, USA
e-mail: jpasari@gmail.com

© Springer International Publishing Switzerland 2016
L.B. Byrne (ed.), *Learner-Centered Teaching Activities for Environmental and Sustainability Studies*, DOI 10.1007/978-3-319-28543-6_15

Learning Outcomes

After completing this activity, students should be able to:

- Summarize, interpret, and critique a scientific article and explain it to others.
- Synthesize findings across scientific articles to identify research gaps.
- Propose hypotheses about Lyme disease ecology.
- Propose Lyme disease mitigation strategies.

Course Context

- Developed for upper- and lower-level ecology and conservation biology courses with 20–30 students. Also applicable to advanced high school students in AP Biology or Environmental Science
- 90 min over two class periods
- No specific background preparation is necessary though the activity is intended for students who are completing or have completed a lower division, introductory unit on ecology
- Adaptable to smaller class sizes by reducing the number of articles read and the number of student groups

Instructor Preparation and Materials

Instructors should possess an understanding of basic Lyme disease ecology, such that they can provide an introductory lecture on the topic (example slides are provided as Electronic Supplementary Materials (ESM) A). Instructors are advised to read Ostfeld (2013) or Ostfeld (2010) for a brief but sufficient synopsis of Lyme disease ecology. For additional background, see the articles listed in ESM-B. In addition, instructors should be well acquainted with the six primary articles used in this activity (Barbour and Fish 1993; Allan et al. 2003; LoGiudice et al. 2003; Ostfeld et al. 2006; Swei et al. 2011; Levi et al. 2012) and be able to discuss them in context of the student worksheet questions (ESM-C). Of these six articles, instructors should decide which ones and how many articles to use in the jigsaw activity (see Activities) depending on class size and student and instructor interest. These articles introduce perhaps the most-discussed and debated hypotheses about Lyme disease spread. They present research conducted at a variety of scales and with a diverse range of methodologies, which allows students to wrestle with results that can seem contradictory until they more closely examine issues of scale. Instructors may also wish to choose studies done in ecosystems similar to their local environment to increase the relevance to students or search the literature for more recently published studies.

Before class, instructors should review the student worksheet (ESM-C) and edit it to match their needs. (For example, instructors may wish to separate part 2 of the worksheet and only hand it out in class when students join their second jigsaw groups so they do not work on it before class.) In addition, the instructor must determine the structure of the jigsaw to assign students to groups. (For more description of the jigsaw method, including a diagram, see Chap. 26, this volume.) A jigsaw activity is structured with two parts. In part 1, students are put in groups to read and discuss one of the six articles that they have all read. In part 2, the groups are reconfigured so that each member has read a different article, and students work together to synthesize information from across them. For example, if there are 24 students in the class, the instructor could assign six unique articles to six 4-person groups or four unique articles to four 6-person groups in part 1. Alternatively, the instructor could assign five unique articles to four 5-person groups and have a fifth group with 4 people. As needed in part 2, groups could have more than one member from a part 1 group.

Activities

After providing a ~15 min introductory lecture on Lyme disease ecology (e.g., ESM-A) and the structure of the activity, the instructor should divide the class into groups, each of which will be assigned an article. Students should be instructed to read their assigned article for homework and answer the questions in part 1 of the handout (ESM-B) to guide their reading.

In the next class meeting, students should first work together in assigned group 1 to review and revise their answers to part 1 of the worksheet (~30 min). The instructor should help guide students and respond to questions by checking in with groups frequently. Some of the articles are technical, and, in the author's experience, students usually have many questions. Encourage students to focus on the results presented in graphs, especially if their article includes unfamiliar math or statistics. Also instruct them to think about how they will convey the most important points about their article to other students in their group 2.

Once each group has finished, ask the students to move to their assigned group 2. Instruct the students to share what they learned in part 1 with the members of their new group. Each student should take only a few minutes to share to ensure that everyone has time to do so (within ~45–60 min). Once each student has shared their article, students complete part 2 of the worksheet by discussing and answering several questions that require synthesis across all of the articles.

The instructor can conclude the activity by allowing each group to share answers to the questions in part 2 and facilitating a class discussion about the similarities and differences between the hypotheses and conclusions proposed by different groups (10–20 min). In particular, the instructor should emphasize the following in context of the Lyme disease debate:

- Conclusions that can be drawn from ecological studies are limited by many factors including: replicability, geographic and spatial scale, sample size, confounding factors, and environmental variation.
- Science involves debate and weighing the strength of seemingly contradictory evidence.
- Synthesis is an ongoing process that requires staying current with emerging scientific literature.
- Critiquing other scientists' methods and conclusions is critical to advancing our understanding.
- Sound scientific debate involves supporting claims with evidence and citations.

Follow-Up Engagement

- The following prompts can help extend students' critical thinking:
  - Design a study that would address an existing research gap and improve our understanding of Lyme disease ecology.
  - How might global changes (e.g., climate change or the spread if invasive species) both promote and hinder the spread of vector-borne infectious diseases like Lyme disease?
  - In what ways is Lyme disease a good case study for examining disease ecology generally? In what ways might it not be generalizable?
- Ask students to update their knowledge by searching for and reading the abstracts of the latest literature to see how the state of the debate is evolving (see ESM-A). Similarly, students can also research the state of research on other vector-borne diseases (e.g., West Nile virus, leishmaniasis). (Such topics could be analyzed using the meta-analysis activity described in Chap. 9, this volume.)
- Ask students to compare their understanding of Lyme disease with its depiction in the popular press (e.g., see Zimmer 2013). (See Chap. 37, this volume, for a relevant activity.)

Connections

- This activity helps students engage in systems thinking (e.g., Chaps. 3, 10 and 31, this volume) because it asks them to assess the relative direct and indirect contributions of many aspects of environmental change (e.g., land cover, biodiversity, climate) to the emergent property of Lyme disease risk.
- The ecology of Lyme disease can be used as a case study for examining complex relationships between the environment and human health and well-being. For

instance, the "dilution effect" described in several of the jigsaw articles connects strongly to the concept of ecosystem services (see Chap. 10, this volume).

- Students often propose Lyme disease mitigation strategies involving land use and wildlife management that intersect strongly with the study of environmental policy and species conservation laws. (See Chap. 20, this volume, for a relevant land management activity.)

References

Allan BF, Keesing F, Ostfeld RS (2003) Effect of forest fragmentation on Lyme disease risk. Conservat Biol 17:267–272

Bacon RM, Kugeler KJ, Mead PS (2008) Surveillance for Lyme disease—United States, 1992–2006. MMWR Surveill Summ 57(10):1–9

Barbour AG, Fish D (1993) The biological and social phenomenon of Lyme-disease. Science 260:1610–1616

Levi T, Kilpatrick AM, Mangel M, Wilmers CC (2012) Deer, predators, and the emergence of Lyme disease. Proc Natl Acad Sci 109:10942–10947

LoGiudice K, Ostfeld RS, Schmidt KA, Keesing F (2003) The ecology of infectious disease: effects of host diversity and community composition on Lyme disease risk. Proc Natl Acad Sci 100:567–571

Ostfeld R (2010) Lyme disease: the ecology of a complex system. Oxford University Press, Oxford

Ostfeld RS (2013) Ecology of Lyme disease. In: Weathers KC, Strayer DL, Likens GE (eds) Fundamentals of ecosystem science. Academic, Waltham, pp 243–251

Ostfeld RS, Canham CD, Oggenfuss K et al (2006) Climate, deer, rodents, and acorns as determinants of variation in Lyme-disease risk. PLoS Biol 4(6):e145. doi:10.1371/journal.pbio.0040145

Swei A, Ostfeld RS, Lane RS, Briggs CJ (2011) Impact of the experimental removal of lizards on Lyme disease risk. Proc Biol Sci 278:2970–2978

Zimmer C (2013) The rise of the tick. Outside magazine. http://www.outsideonline.com/1915071/rise-tick. Accessed 21 Aug 2015

Chapter 16
A Fishery Activity to Examine the Tragedy of the Common Goldfish Cracker

Jessa Madosky

Introduction

"The tragedy of the commons" is often referenced in discussions of natural resource management and other environmental issues. Hardin (1968) argued that common resources are bound to be overexploited unless exploitation is limited by government or private ownership and that these problems are not solvable through scientific advancement or understanding alone. However, Ostrom (1998) identified collective action involving trust, reciprocity, and communication as an alternative path for managing common resources. Recent work suggests that Hardin's conclusions may hold true for large-scale (national, regional, or global) commons, while Ostrom's work may be more useful in small-scale commons (Araral 2014). Fisheries, commonly used as an example of the tragedy of the commons, have been found to exhibit these outcomes such that shared fish stocks often perform more poorly than individually owned ones and large-scale management of shared stocks is more likely to lead to overexploitation (McWhinnie 2009; Burger and Gochfeld 1998).

In the exercise described in this chapter, students play the role of fishers to simulate the tragedy of the commons using goldfish crackers. Working in groups, students should learn about population dynamics and discover the challenges of sustainably managing a "public" natural resource with natural variation and the potential conflict between self-interest and the public/ecological good. Although this exercise has been proposed in various forms for a high school audience (see Electronic Supplementary Materials (ESM)-A), the version described here for

Electronic supplementary materials: The online version of this chapter (doi:10.1007/978-3-319-28543-6_16) contains supplementary material, which is available to authorized users.

J. Madosky (✉)
Biology Department, University of Tampa, Tampa, FL, USA
e-mail: jmadosky@ut.edu

© Springer International Publishing Switzerland 2016
L.B. Byrne (ed.), *Learner-Centered Teaching Activities for Environmental and Sustainability Studies*, DOI 10.1007/978-3-319-28543-6_16

undergraduates increases the complexity of the exercise by including ecological variability in the form of variable growth rates from year to year and Allee effects and by allowing students to purchase data to guide fishery management or equipment to increase their harvest. Overall, the goal is for students to understand "the tragedy of the commons" and why common resources can be so difficult to manage, especially in the face of environmental variation.

Learning Outcomes

After completing this activity, students should be able to:

- Describe the concept of "the tragedy of the commons" and apply it to natural resource management.
- Discuss challenges in managing a common resource.
- Create and assess management plans for a fishery and discuss the trade-offs between satisfying humans' resource needs and ensuring a viable, long-term fishery.
- Explain why some management problems may not be solvable with more scientific knowledge.

Course Context

- Developed for an upper-level conservation biology course for science majors with 15–20 students placed into groups of 4–5
- 75–105 min in 1 or 2 class sessions
- No background knowledge is required, but students can prepare by reading Hardin (1968)
- Can be adapted for larger groups (additional assistance counting crackers may be necessary) and for any level of course that would cover natural resources and can also be adjusted for a shorter duration by only using Part 2 of the exercise or reducing fishing rounds or for a longer duration by extending discussion or fishing rounds. Variations for different student levels are provided in the supplementary materials

Instructor Preparation and Materials

To prepare, the instructor may wish to read Hardin (1968), Araral (2014), McWhinnie (2009), and Burger and Gochfeld (1998) for scientific background of some of the key concepts of "the tragedy of the commons" and fishery management. In addition, news articles about the current state of fisheries can provide helpful context for discussions (see ESM-B for more background readings). The instructor may also wish to assign Hardin (1968) as reading prior to or after the exercise; students can be successful in this exercise without reading it, and the instructor may prefer for students to learn the concept actively before reading about it.

Before the activity, the instructor will need to obtain goldfish crackers (approximately 1000 per group of four or five students), plastic spoons (two per student), cups or bowls (one per student plus one for each group that is labeled with a group number), and a typical playing die. Forks or knives could be used instead of (or with) spoons to make fishing harder and simulate different types of fishing equipment, and metal utensils could reduce risks of breakage.

To create the fisheries, place 200 goldfish into each of two cups (one cup for Part 1 and one cup for Part 2) for every group of four or five students. The instructor can form student groups before or at the beginning of class in a manner desired but the exercise works best if students do not work with their close friends to more accurately simulate competition with "strangers" in a large fishery. Students can perform the simulation on a table or the floor; the space needs to be large enough that all students can gather around the "ocean" containing the fish simultaneously.

The instructor should also download, and prepare as relevant, the set of supplementary files useful for implementing the game. These files include detailed instructions and an example script (ESM-C) and additional discussion questions and prompts (ESM-D) instructors may find useful in implementing the exercise and which can be modified to suit instructor preference. The files also include additional variations on the exercise that may be implemented in addition to other versions of the exercise previously published and additional references and reading materials (ESM-A). The instructor should print or display the fishery rules and price list for each group (ESM-F) or display it prominently in the classroom so all students can see these rules and prices as they conduct the exercise and print a student data sheet for each student or each group (ESM-G) to facilitate data collection throughout the exercise. The instructor may also wish to print the instructor price guide and environmental factor key (ESM-E) for themselves to use as a reference during the exercise.

Activities

The activity is implemented in two parts. In Part 1, students conduct the fishery exercise without any instruction to sustainably manage their fish stock. Then in Part 2, students conduct the same fishery exercise while attempting to sustainably manage their fish stock. A brief overview of the activity is presented below; see ESM-C for detailed instructions, an example script for running the exercise, and discussion questions and prompts to share with students.

Part 1: Unguided Fishery Exploitation (~30–45 min depending on the number of fishing rounds)

Introduction

- Prepare students for the activity by briefly introducing the simulation (a sample script is provided in ESM-C).
- Have groups gather around an "ocean" space (table or floor) and provide each student with a cup or bowl (their processing facility) and a plastic spoon (their fishing vessel).

First Round

- Place 200 goldfish crackers in each group's ocean (spreading out the goldfish makes it harder to catch them).
- Allow students to fish for 30 s, reminding them that they need to catch at least eight fish to "survive" to the next round.
- Collect the uncaught fish in the cup labeled with that group's number.
- Students should count and record the number of fish they caught on a data sheet (ESM-B).
- Student that caught eight or more fish made it through this fishing season successfully. They should eat or hand in their eight fish to represent the fish needed for them to feed their family for the year and stay in business. Students may use additional caught fish to make any purchases they desire (giving the fish to the instructor as payment; see ESM-E) or save them for later to offset a bad season.
- Student with fewer than eight crackers are "bankrupt" and must sit out the next round unless a fellow fisher is willing to hire them.
- If a group has depleted their fishery (by catching all of their fish), they will not have any to catch the next season. They may try to find jobs fishing for someone in another group.

Prepare for Second Round

- Allow students to talk undirected while the instructor (and/or an assistant) prepares for the second round—at this point it is preferable for students to talk unguided to simulate what would occur in a fishery where the fishers were not actively managing the fish. Students often talk about the number of fish caught, but the instructor should avoid guiding students to reflect on the exercise.
- Roll the die (out of the students' sight) to determine if it is a "good", "bad," or "average" year in terms of environmental conditions for the fish's survival and reproduction (see ESM-E).
- Count out the starting population for each group based on the fish left in the population and the die roll (based on guidelines for calculations provided in ESM-E).
- If any fisher or group has purchased a study, provide them with the information purchased (what type of year it is based on the die roll, growth rates for different types of years, or number of fish in their population after fishing; see ESM-E).

Only the purchaser(s) should be given this information, but the purchaser can share the information if desired.

Second Round of Fishing

- Give each group its cup of fish which are placed back into the group's ocean.
- Start the next 30 s season of fishing.
- Follow the same steps as above for round 1.

Subsequent Rounds

- Repeat the above steps for additional rounds of fishing until several of the groups have depleted their fish populations, usually after four or five rounds, or until several of the groups have worked out strategies to maintain their fish populations.

Discussion

- Ask students to report back to the whole class on what happened to their fish population and to them as fishers and why (also see questions in ESM-D). If some students have already developed strategies to manage and maintain their populations, ask them to explain their approaches.

Part 2: Working Toward Sustainable Fishery Management (~45–60 min depending on the number of fishing rounds and time for discussion between rounds) This part can be conducted in the same class period if time permits or in a following one. If the latter, the instructor may want to ask students to develop a management strategy for homework.

Introduction

- Inform the class that this time they will be trying to manage their fish population to ensure a sustainable population but that there are no "rules" on how they should do this. They will be creating the management strategy and any regulations as they go.

First Round and Preparation for the Next Round

- Restart the fish population at 200 and conduct the session in the same way as Part 1 above, but in between fishing rounds, ask students to discuss, as a group, how they will manage their fish population, for example, by creating (or changing) regulations.

Subsequent Rounds

- Continue with additional fishing rounds until most groups' populations either become somewhat stable or crash. The number of rounds needed for these outcomes will vary with die rolls and group dynamics but may be as few as three rounds or as many as nine rounds. If time is limited, the instructor may want to limit it to four or five rounds to save time for discussion.
- The instructor may also want to limit between-round discussion time to leave more time for follow-up discussion.

Discussion

- Ask students to report back to the whole class on what happened to their fish population and to them as fishers when they managed the population and how they managed their population.
- Depending on available time, discussion questions could be used for in-class dialogue and/or for homework assignments (see ESM-D for questions about environmental variability, technological solutions, and connections to historical fishery collapses and Hardin (1968)).

Follow-Up Engagement

- Further assignments could include a short essay relating the in-class fishery exercise to Hardin's (1968) article (prompt ideas are provided in ESM-D). If more in-depth discussion of managing common resources is desired, Araral (2014) may be assigned to inform discussion on differing views of how to manage common resources.
- Students could be asked to research real fisheries and either write about or briefly report to the class on the management plans implemented and any successes or challenges in real-life fisheries.
- For a related activity about lake food webs that includes an overfishing scenario, see Chap. 13, this volume.

Connections

Suggested assignment prompts for the following connections are provided in ESM-D.

- This exercise directly connects to ecological concepts such as carrying capacity, population growth rates, and how stochastic environmental variation can impact those variables.
- When connected with real-life fishery management, the exercise can be used to introduce sustainable use, maximum sustainable yield, total allowable catch, or other management strategies commonly used in fishery management.
- More broadly, this exercise can be used to introduce the role of economic and ethical arguments in conservation and resource management and launch discussions about human impacts on seemingly limitless resources. (For related activities, see Chaps. 17–20, this volume.)
- This exercise also links well to discussing the role of politics in environmental decisions and can be used to introduce the concept of stakeholder interests and diverse views (e.g., Chaps. 4, 27, 29 and 39, this volume). It can also prompt discussion about public distrust of science and the issues with "top-down" mandates.

References

Araral E (2014) Ostrom, Hardin and the commons: a critical appreciation and a revisionist view. Environ Sci Pol 36:11–23

Burger J, Gochfeld M (1998) The tragedy of the commons 30 years later. Environment 40:4–13

Hardin G (1968) The tragedy of the commons. Science 162:1243–1248

McWhinnie SF (2009) The tragedy of the commons in international fisheries: an empirical examination. J Environ Econ Manag 57:321–333

Ostrom E (1998) A behavioral approach to the rational choice theory of collective action: presidential address, American Political Science Association, 1997. Am Polit Sci Rev 92:1–22

Chapter 17
Making Biodiversity Stewardship Tangible Using a Place-Based Approach

Christine Vatovec

Introduction

Environmental stewardship is a concept heralded by Aldo Leopold (1949) as the responsible use and conservation of natural resources in a way that protects and sustains these resources. It is a management concept that responds to the impacts that human activities have on ecological systems, and biodiversity in particular. Chapin et al. (2011) call for stewardship approaches that recognize the integral relationships between humans and the environment within social-ecological systems (see Box 17.1). Humans rely on biodiversity to provide food, fibers, and other materials that support our basic needs and well-being. Therefore, to maintain healthy human communities we must take actions to conserve the biodiversity on which we depend (Frumkin 2010). For example, agricultural practices that favor large-scale monoculture crops could lead to human starvation if the crops are decimated by a natural disaster (drought, pest outbreak, etc.) to which they have no resistance. In contrast, a more biodiverse agricultural system could be more resilient to such disasters, thus better supporting human well-being. Recent research on the relationship between biodiversity and agricultural pest outbreaks highlights the importance of such interconnections (Lundgren and Fausti 2015).

The class activity described in this chapter provides an opportunity for students to apply the concept of stewardship in a tangible way to the biodiversity of a place in which they have a personal interest. It is framed by the *One Cubic Foot* project conceived of by photographer David Liitschwager (2012) who documented the biodiversity of various ecosystems around the world that resided in or passed

Electronic supplementary materials: The online version of this chapter (doi:10.1007/978-3-319-28543-6_17) contains supplementary material, which is available to authorized users.

C. Vatovec (✉)
Rubenstein School of Environment & Natural Resources, University of Vermont, Burlington, VT, USA
e-mail: cvatovec@uvm.edu

Box 17.1. Types of Stewardship Actions (Also See Slides in ESM-C)

The Ecosystem Stewardship framework by Chapin et al. (2010) identifies three primary strategies for sustainably managing biodiversity and thereby enhancing human well-being in social-ecological systems. The three action-oriented strategies include (with examples for agricultural systems):

1. Reducing known stressors that negatively impact biodiversity (e.g., reducing grazing pressure to sustain ecosystem well-being in drylands),
2. Developing proactive adaptive management strategies when change is uncertain (e.g., incentivizing mixed-cropping systems over monocultures to buffer crop failures and enhance economic vitality), and
3. Transitioning away from social-ecological systems that are trapped in producing undesirable outcomes (e.g., shift away from large-scale monoculture agriculture toward agro-ecological landscape management).

through a one-cubic foot metal frame over a 24-h period. In Costa Rica, for example, he found more than 150 species of plants and animals in the cube. Craig Childs (2012) extended the project to examine the impacts of human activities on biodiversity in a conventionally managed Iowa cornfield where he found only seven species in addition to the corn. In the context of the cornfield, biodiversity stewardship actions may include maintaining crop diversity, creating habitat for pollinators, employing cover crops, and using integrated pest management, all of which could help conserve and enhance biodiversity (FAO 2015), and may better support human well-being by accounting for the interconnections of this social-ecological system.

The exercise asks students to extend their insights from this agricultural example to explore human activities that could support the biodiversity of a place of their choosing. Students may choose to examine any location they have visited, from their own backyard, to a parking lot, to a national park. Through in-class discussions, students should develop a broader awareness of stewardship as a tool of sustainability for conserving and increasing biodiversity. The goal of this learning activity is for students to deepen their understanding of "stewardship" in terms of sustainability in interconnected social-ecological systems by thinking about how human actions affect biodiversity in both positive and negative ways.

Learning Outcomes

After completing this activity, students should be able to:

- Describe how human activities impact biodiversity.
- Discuss the role of environmental stewardship in sustainability.
- Reflect on various human actions to determine which may be best implemented for stewarding biodiversity in different ecosystems.

Course Context

- Originally implemented in an upper-level undergraduate course in sustainability science with an enrollment of 30–40 students
- 50 to 90 min in one class meeting
- Before class, students read two articles and complete an assignment; instructors may choose to introduce the topic of biodiversity stewardship using the slides provided (ESM-C)
- Adaptable to courses of different class sizes and levels that include discussions of biodiversity conservation; can be modified for longer durations by extending the activity to take place over multiple class sessions

Instructor Preparation and Materials

To successfully help students achieve the learning outcomes, instructors should be able to (a) describe the concept of stewardship in the context of sustainability, (b) describe the interconnections between social and ecological systems (for example, how food choices relate to agricultural practices, which in turn affect biodiversity in agricultural systems), and (c) explain the relationship between biodiversity and human well-being. The concept of stewardship as a tool for moving toward sustainability via a systems approach to understanding interconnections between social processes (i.e., human activities) and ecological outcomes has been described in detail as a part of the Ecological Society of America's "Earth Stewardship" initiative (ESA n.d.; Chapin et al. 2010, 2011). The connections between biodiversity and human well-being, for example the relationship between levels of biodiversity and quality of air, water, soil, and pathogen regulation, are described in Frumkin (2010: p. 25) (also see Chaps. 10, 11, 13 and 15 for more references and context).

To complete the activity, instructors should read the two articles that will be assigned to students (Wilson 2010; Krulwich 2012) and review, edit, as desired, and print the assignment sheets to hand out in a class prior to the activity (provided in Electronic Supplementary Materials (ESM) A). Instructors may also benefit from reviewing the example student response (ESM-B). During class, instructors need access to an overhead projector or large sheets of paper for student groups to share their insights from group discussions, and a chalk or white board or electronic tools to record notes from the discussion. Instructors may also use the provided presentation slides to introduce the activity (see ESM-C).

Activities

Prior to class, the instructor will assign students the following tasks (see ESM-A):

1. Visit National Geographic's "A cubic foot" project website to read article by Wilson (2010) and review the photo gallery;

2. Read Krulwich (2012);
3. Prepare a one page, single-sided document to share in class that includes a pho-
   tograph of a place of personal interest to the student, and a short description of
   one stewardship action that would increase the number of species in that place,
   as modeled after the *One Cubic Foot* project.

In class, the instructor can introduce the topic of stewardship as best fits within
the course before giving students the homework assignment (see resources in the
previous section as a guide). On the day the homework is due, the instructor asks
students to form groups of four. Each group sits in a circle, and students pass around
their prepared sheets while noting themes/patterns and outliers seen in their peers'
suggested stewardship actions. (The instructor can indicate the time (e.g., every
45–60 s) when the images should be passed to the next person.) After students have
reviewed all their group members' sheets (~4 min), students in the group pair up and
compare notes on the themes and outliers they observed (~4 min). Then the group
of four should discuss their observations and develop a set of general biodiversity
stewardship actions to present to the whole class, including an overview of interconnections between human actions and biodiversity (~15 min). Each group then
selects a presenter to share their group's findings with the rest of the class (e.g., on
a chalk or whiteboard; ~3 min per group). Students should be asked to note the similarities and differences among the groups' findings, and then develop a list of the
different types of actions that may increase biodiversity in the various settings,
using the framework described in Box 17.1 (~5–10 min). The instructor then guides
the students in a discussion of interconnections between stewardship actions and
possible biodiversity outcomes in the various places/ecosystems represented (~5–
10 min). Here it may be useful to select a few of the students' assignments to share
on a projector to compare the different types of stewardship actions that may be best
suited for different places while referring to the three types of stewardship actions
described in Box 17.1.

Follow-Up Engagement

- Assign students a short writing assignment to reflect upon their class discussions
  and to synthesize ideas about biodiversity stewardship in different contexts.
- Ask students to discuss why the place they chose is the way it is and what barriers they might face when trying to steward it. Also, ask them to apply the
  social-ecological systems framework to analyzing the characteristics of the place
  (e.g., see Chap. 31).
- Prompt a discussion with students about how Leopold's (1949) definition of
  stewardship, the responsible use and conservation of natural resources in a way
  that protects and sustains these resources, relates to their chosen place.
- For activities related to conservation and land use planning, see Chaps. 18–20.

Connections

- If this activity is completed early in the course, the instructor could revisit it later to critically examine earlier assumptions and to further develop students' understanding of interconnections between human actions and ecological outcomes.
- When discussing case studies that present imperiled ecosystems, compare different stewardship actions that could work to resolve the problem and move the system toward a more sustainable outcome in which biodiversity is conserved.

References

(ESA) Ecological Society of America. (n.d.) Earth stewardship. http://www.esa.org/esa/science/earth-stewardship/. Accessed 20 Oct 2014

(FAO) Food and Agriculture Organization of the United Nations (2015) How to manage agricultural biodiversity. http://www.fao.org/agriculture/crops/thematic-sitemap/theme/spi/scpi-home/managing-ecosystems/biodiversity-and-ecosystem-services/bio-how/en/. Accessed 29 June 2015

Chapin III FS, Power ME, Pickett ST et al (2011) Earth stewardship: science for action to sustain the human-earth system. Ecosphere 2.8: art89

Chapin FS III, Carpenter SR, Kofinas GP et al (2010) Ecosystem stewardship: sustainability strategies for a rapidly changing planet. Trends Ecol Evol 25:241–249

Childs C (2012) Apocalyptic planet: field guide to the future of the earth. Pantheon, New York

Frumkin H (ed) (2010) Environmental health: from global to local. Wiley, New York

Krulwich R (2012) Cornstalks everywhere but nothing else, not even a bee. National Public Radio. http://www.npr.org/blogs/krulwich/2012/11/29/166156242/cornstalks-everywhere-but-nothing-else-not-even-a-bee. Accessed 10 Oct 2014

Leopold A (1949) A Sand County Almanac. Oxford University Press, New York

Liitschwager D (2012) A world in one cubic foot. University of Chicago Press, Chicago, IL

Lundgren J, Fausti S (2015) As biodiversity declines on corn farms, pest problems grow. The conversation. http://theconversation.com/as-biodiversity-declines-on-corn-farms-pest-problems-grow-45477. Accessed 15 Aug 2015

Wilson EO (2010) A cubic foot. National Geographic. http://ngm.nationalgeographic.com/2010/02/cubic-foot/wilson-text. Accessed 10 Oct 2014

Chapter 18
Conservation Triage: Debating Which Species to Save and Why

Loren B. Byrne

Introduction

The goal of conservation biology, as both science and practice, is to prevent species from going extinct. Unfortunately, so many species are now threatened with extinction that it may not be possible to save them all, especially in context of limited and sometimes-declining resources (Marris 2007; Bottrill et al. 2008). This situation presents difficult scientific and philosophical questions about which species should be prioritized for conservation and why. On the flip side the question is, which species should we not invest in, thus increasing the risk of their extinction (Jachowski and Kesler 2009; Groc 2011)? As one journalist queried, "Which species will you save? Will you pick the rarest, the largest, or the smallest? The strongest or the weakest? The most beautiful ... or just the tastiest?" (Nijhuis 2013). Akin to triage in a medical emergency situation, these issues have brought forth the idea of conservation triage: having to make difficult choices about which species should be saved and which might not be savable.

The activity described in this chapter engages students in a structured debate about conservation triage (that includes "triage emergencies" and votes about which species to save) to help them learn about diverse endangered species and the factors influencing their population declines and conservation status. The activity's broader goal is to foster students' critical thinking skills by engaging them in a discussion about complex issues that mirrors one within the professional conservation community. As they share and debate arguments to justify and prioritize conserving

Electronic supplementary materials: The online version of this chapter (doi:10.1007/978-3-319-28543-6_18) contains supplementary material, which is available to authorized users.

L.B. Byrne (✉)
Department of Biology, Marine Biology & Environmental Science, Roger Williams University, Bristol, RI, USA
e-mail: lbyrne@rwu.edu

© Springer International Publishing Switzerland 2016
L.B. Byrne (ed.), *Learner-Centered Teaching Activities for Environmental and Sustainability Studies*, DOI 10.1007/978-3-319-28543-6_18

different species, they will be encouraged to reflect on the challenges of biodiversity conservation and examine the roles of evidence and criteria for making difficult environmental and sustainability decisions.

Learning Outcomes

After completing this activity, students should be able to:

- Describe various factors that affect species' endangerment and conservation status.
- Define triage and apply the concept to examining sets of endangered species.
- Discuss the information needed and factors to consider when making conservation triage decisions.
- Articulate personal views about the value of different endangered species.
- Explain how ethics and personal values relate to the study and practice of conservation biology.

Course Context

- Developed for an upper-level conservation biology course for science majors with 20–30 students
- 50 min in one class meeting
- Before class, students gather information to present in class but no additional background knowledge is needed
- Adaptable to courses of any class size and level that include discussions of biodiversity conservation; can be modified for longer or shorter durations by altering the number of species and student groups

Instructor Preparation and Materials

Instructors should be familiar with issues of conservation triage which can be gained by consulting the references cited in the introduction; additional perspectives can be found in Kareiva and Levin (2003). In addition, a source for creating a list of endangered species to include in the debate is needed. The one that inspired this activity is the book "100 Under 100: The Race to Save the World's Rarest Living Things" by Leslie (2014). However, other compilations of endangered species could be consulted to derive focal species to include in the activity (e.g., Goodall 2009; Red List n.d.); an online search with the phrase "100 most endangered species" provides many resources, including a Wikipedia article. Alternatively, instructors could invite students to choose their own endangered species to use.

Prior to the in-class activity, the instructor creates a list of endangered species that will be subjected to triage decisions by the students. For the activity to work best, it is recommended that the list of species include a range of diverse species, including ones from different taxa (e.g., mammals, birds, invertebrates, plants) and ones that vary in their popular appeal (i.e., more charismatic ones like primates, other cute mammals and "pretty" species like butterflies alongside less well-known and "ugly" species). (An example set meeting these criteria, all included in Leslie (2014), is presented in Table 18.1.) The number of chosen species depends on the class size and number of students assigned to each species. Each species will be assigned to at least two students; one student will argue in favor of prioritizing the species for conservation efforts ("save it") and the other will argue that it can be "let go" (i.e., not receive conservation priority). For larger classes, more than two students could be assigned to each argument for each species. Alternatively, a third student could be assigned to take a neutral position for a species and serve as an impartial referee when each species is presented in class. Based on the list of focal species, the instructor assigns each student a species and triage argument (for or against the species receiving conservation priority, or neutral as needed) sufficiently in advance of the class meeting when the activity will occur to allow students time to complete the preparatory assignment described below. As desired, to help students complete the background research for their assigned species needed for the in-class debate, an assignment sheet can be created (tailored to the course and student context) with guiding questions and resources (e.g., Leslie 2014; Red List n.d.); in the author's experience, upper-level science majors students were able to successfully complete this task when given simple oral instructions to "prepare an argument, with evidence and justification, for why your assigned species should or should not be conserved."

In addition, the instructor may wish to create a set of presentation slides with images of the chosen species to be displayed as each species is discussed by the students. (Example slides for the list of species in Table 18.1 is provided as

Table 18.1 Example set of species to include in the triage debate

Common name	Scientific name
Alabama sturgeon	*Scaphirhynchus suttkusi*
Arakan forest turtle	*Heosemys depressa*
Armoured mistfrog	*Litoria lorica*
Catalina mahogany	*Cercocarpus traskiae*
Cat Ba langur	*Trachypithecus poliocephalus*
Eastern North Pacific right whale	*Eubalaena japonica*
Franklin's bumblebee	*Bombus franklini*
Hainan gibbon	*Nomascus hainanus*
Hawaiian crow	*Corvus hawaiiensis*
Javan rhinoceros	*Rhinoceros sondaicus*
Northern river shark	*Glyphis garricki*
Seychelles sheath-tailed bat	*Coleura seychellensis*
Thermal water lily	*Nymphaea thermarum*

Electronic Supplementary Materials (ESM) A.) Among the species slides, "emergency triage" slides can be inserted after every two or several species to indicate moments when the class will collectively decide (via voting) which species to save and which to let go (see the structure for this in ESM-A). The sequence of the species presentations should be organized so as to present easier and then successively harder triage decisions for a few pairs and, then, groups of species. For example, using the species in Table 18.1, when deciding to vote for the Catalina mahogany (a non-descript shrub endemic to one island) or the Cat Ba langur (a fuzzy and cute primate with adorable orange babies), most students will probably easily choose to prioritize the langur. However, deciding between the Javan rhinoceros and the armoured mistfrog may be less straightforward and engender more personal conflict and passionate debate among students. Calling for votes among three or more species increases the challenge level, requires more thoughtful discussion, and reflects the real-world debate.

Activities

Before the in-class debate, students investigate some of the basic biology and conservation status of their assigned species, using for example Leslie (2014), other key references provided by the instructor (e.g., Goodall 2009; Red List n.d.) and/or through independent online searches. (If all the assigned species are from Leslie (2014), it could be provided for students to access in closed reserve in a library; alternatively, pages relevant to chosen species can be copied and given to the students.) In particular, students should look for information that relates to the feasibility of conserving the species in perpetuity, choosing to highlight factors that relate to their assigned triage position (e.g., a species does not breed well in captivity so it should be let go; the population is increasing so it should not be abandoned). The instructor may wish to require that the students complete this as a written assignment and turn something in for a score or participation credit.

In the classroom, the instructor can introduce the concept of triage and other context as desired before the activity (e.g., see a few context slides in ESM-A). Then, in turn, each pair (or more) of students presents the pro and con views for their respective species, essentially justifying to their classmates why their assigned species should be conserved or not. The duration of each presentation and whether or not to include follow up discussion is at the discretion of the instructor based on the class length and duration.

To foster tough decisions, the instructor should intermittently—and without warning to cultivate a sense of surprise, just like in the real world—call for an "emergency triage vote" about which of the previously presented species should be saved—because not all of them can! (See the flow of slides in ESM-A for how to structure the debate and include the emergency slides.) Allowing all students to briefly advocate and debate their triage decisions before each vote is taken will help contribute to an energized and engaging classroom dynamic.

With each successive emergency triage vote, the original set of assigned species will be whittled down to the "winners" of each triage decision. At the end of the presentations, a final triage vote can be requested to identify the final winner: the one species from the original list that will receive conservation prioritization. Although this may be a tough reality for students to consider—that only one of many endangered species might be effectively conserved—forcing such a decision reflects the reality that conservation organizations must face all the time, i.e., where to allocate scarce resources. To wrap up the debate, ask students to reflect on, summarize and synthesize the information (including from the student presentations) and factors (including their personal biases) that influenced their triage decisions and what other considerations (e.g., political, legal, and economic issues) might need to be considered when triaging species. To ensure all students' engagement, this could be done with a "minute paper" in which students write down their personal reflections.

Follow-Up Engagement

- The following questions can be used to extend the discussion:
    - What types of species do we care most about conserving and why?
    - What is the value of an individual species? Are some more valuable than others?
    - What are the costs (financial, social) and trade-offs of conserving species? If the downsides are too great, is it ever ok to "give up" on a species? Under what conditions might it be best to do so?
    - When a species goes extinct, does anything really change? If not, is it ok to let species go extinct?
    - If species can be brought back from extinction (de-extinction), does that support the argument for letting them go?
- Assign students to write an opinion essay that summarizes their views about biodiversity conservation generally or triage decisions specifically.
- Ask students to investigate an in-depth species conservation case study that involves social and political conflicts (e.g., the grey wolf in the Western USA) and highlights the challenges of, yet need for, engaging local communities.

Connections

- Triage can be connected to lessons about the causes of species endangerment and the need to consider larger-scale conservation issues such as ecosystem preservation and land use management (e.g., see Chaps. 12, 16, 17, 19 and 20).
- Relate this activity to themes of ecosystem services (e.g., Chap. 10) and economics; should we prioritize species that provide tangible benefits to humans?

- People's worldviews (e.g., Chap. 4) and environmental ethics are likely to affect how they view species conservation. Some people mock attempts to conserve certain species like endangered beetles and flies.
- Issues of politics and conservation laws (e.g., US Endangered Species Act and international CITES treaty) strongly influence conservation decisions and the ability of conservationists to prevent extinctions.

References

(Red List) International Union for Conservation of Nature and Natural Resources. The IUCN red list of threatened species. http://www.iucnredlist.org/. Accessed 10 Mar 2014

Bottrill MC, Joseph LN, Carwardine J et al (2008) Is conservation triage just smart decision making? Trends Ecol Evol 23:649–654

Goodall J (2009) Hope for animals and their world: how endangered species are being rescued from the brink. Grand Central Publishing, New York

Groc I (2011) Should conservationists allow some species to die out? Discover, March. http://discovermagazine.com/2011/mar/10-should-allow-some-species-die-out#.UyC0cIXDWSo. Accessed 10 Mar 2014

Jachowski DS, Kesler DC (2009) Allowing extinction: should we let species go? Trends Ecol Evol 24:183–184

Kareiva PM, Levin SA (eds) (2003) The importance of species: perspectives on expendability and triage. Princeton University Press, Princeton

Leslie S (2014) 100 Under 100: the race to save the world's rarest living things. Harper Collins, Canada

Marris E (2007) What to let go. Nature 450:152–155

Nijhuis M (2013) Conservation triage. Slate Magazine. http://www.slate.com/articles/health_and_science/animal_forecast/2013/02/conservation_triage_which_species_should_be_saved.html. Accessed 10 Mar 2014

Chapter 19
Everything Cannot Be Equal: Ranking Priorities and Revealing Worldviews to Guide Watershed Management

Bethany B. Cutts

Introduction

Water is an integrating natural resource that connects ecosystems, economies, and societies. In large watersheds like the Upper Mississippi River Basin (UMRB, see Box 19.1), the mismatch between environmental systems and political boundaries means that finding ways to work together is essential to make progress toward identifying shared water management goals and taking action toward them. In the case of the UMRB, for example, we might ask: How do we do uncover and discuss divergent worldviews about how to manage water and what to manage it for, and how do we manage the watershed to improve sustainability for all important uses? Answering these questions requires tools from the social sciences that can help identify and compare stakeholders' different perspectives and priorities for multiple desirable outcomes for managing complex resources. Understanding how people weigh competing outcomes provides information on potential trade-offs while uncovering their rationales can reveal common ground that might otherwise be overlooked. The process of ranking disparate ideas and discussing choices presents an opportunity to initiate critical but difficult discussions about the values that influence how stakeholders view "the facts" as they advocate for their choices for watershed planning and water management.

In the activity described below, students sort, rank, and explain their priorities for managing a watershed using a social science method called Q-sort. This pro-

Electronic supplementary materials: The online version of this chapter (doi:10.1007/978-3-319-28543-6_19) contains supplementary material, which is available to authorized users.

B.B. Cutts (✉)
Department of Natural Resources and Environmental Sciences, University of Illinois at Urbana-Champaign, Urbana, IL, USA
e-mail: bcutts@illinois.edu

© Springer International Publishing Switzerland 2016
L.B. Byrne (ed.), *Learner-Centered Teaching Activities for Environmental and Sustainability Studies*, DOI 10.1007/978-3-319-28543-6_19

Box 19.1

The Upper Mississippi River system is an important waterway that stretches from Lake Itasca to Cairo, Illinois and touches 60 counties in Illinois, Minnesota, Missouri, Iowa, and Wisconsin. It contains a 1300-mile river system that is home to over 154 fish species and an essential migratory pathway for 40 % of American waterfowl and shorebirds (Delong 2005; USGS 2014). In addition to its ecological importance, this river system provides water for industrial uses such as wastewater management and commercial navigation (USGS 2014). The Upper Mississippi River system also plays a key role in the transportation of commodities to and from the Gulf States while providing water to over 30 million residents for drinking uses and daily industrial purposes (USGS 2014). For additional background, see references in ESM-A.

cess reveals similarities and differences among the students' own understandings of sustainability priorities, which can lead to articulating their operational definitions of sustainability. The goal of this learning activity is for students to uncover the roles that their own beliefs, experiences, and motivations play in shaping their perspectives about what *should* be done. Revealed differences among students' ideas replicate those that may be involved in negotiating social sustainability challenges (e.g., Webler et al. 2009; Danielson et al. 2009). Students should discover that, even with "all the facts," working toward sustainability involves negotiating conflicting worldviews (e.g., Chap. 4). The activity should also develop students' normative and interpersonal competencies, both essential to sustainability pedagogy (Wiek et al. 2011).

Learning Outcomes

After completing this activity, students should be able to:

- Identify the (parts of) worldviews of others and how those relate to their interpretations and views of sustainability.
- Recognize ways that values, beliefs, and experiences shape how individuals interpret and organize information to form priorities.
- Explain why stakeholders committed to the same goal prioritize different actions.
- Articulate their own views and personal definition of sustainability (in the water resource context).

Course Context

- Developed for a non-majors environmental studies course; also used in a conservation biology course and during a summer research experience orientation (all with 20–50 students)
- 50 to 90 min in one class meeting
- No student preparation required (e.g., not essential that students receive formal instruction on definitions of sustainability or the technical details and political concerns of the watershed prior to the activity)
- Adaptable to most learning environments and student populations

Instructor Preparation and Materials

Instructors should become familiar with the Q-sort method so that they can introduce the activity and guide students through it. Background and instructions are provided in Electronic Supplementary Materials (ESM) A. For additional context, Meo et al. (2002) and Previte et al. (2007) provide examples of Q-sort applications to environmental policy and water management.

This lesson was developed for the Upper Mississippi River Basin. Instructors may choose to use this case study or adapt the materials to a watershed relevant to their course and/or familiar to their students. Instructors who use the provided Q-sort statements should be familiar with the Upper Mississippi River Basin geography (Box 19.1) and the terminology used in the statements that students will sort (ESM-B). Instructors who wish to adapt the activity to a different watershed will need to prepare 30 statements that relate to sustainability concerns in their region. These statements should represent diverse interests and actions, and represent a variety of issues (e.g., protecting native species; supporting rural livelihoods; reducing toxin loads), including both commonly discussed ones (e.g., reducing nutrient pollution from agriculture to improve water quality; regulating fracking to prevent groundwater contamination) and lesser known ones (e.g., scaling back regulation to allow free-market principals to distribute environmental benefits; limiting noise pollution from shipping barges). As a third alternative, students can develop statements during a preparatory assignment (Box 19.2).

The instructor will need to print out materials for the student sorting activity. Each student will need a set of 30 statements (e.g., ESM-B), a sorting pyramid (ESM-C, printed on 11×17 in. paper at minimum), and the reflection questions (ESM-D). Cut out the statements so each one will fit into a square on the sorting pyramid. The instructor can provide students with scissors and a single sheet containing all the statements; then students can cut out each statement as they read them (during step 4 of "Activities" below). Alternatively, to shorten the time needed for in-class work, the instructor could ask students to read and cut out the statements as homework before the class, or provide each student with an envelope of statements already cut.

Box 19.2

To encourage students to think more deeply about definitions of sustainability in watersheds or other large, multi-jurisdictional planning areas, instructors can ask students to complete a photo essay assignment as homework 1–2 weeks before using the Q-sort activity. In the assignment, students will research sustainability issues for the assigned topic and document 15 related actions, situations, or conditions in photographs. This activity can help students identify the difficulties and complexity of creating an operational definition of sustainability. It also serves as an opportunity to assess student knowledge about the watershed, and provoke brainstorming and reflection. By doing it *before* the Q-sort, students may begin to notice biological, social, and economic interests in the watershed that they had not previously considered.

Activities

The text below provides an overview of steps for the activity. Additional details and background, including suggested instructions for students, are included in ESM-A.

1. Introduction (3–5 min). Introduce the lesson to students, explaining that the goals of this activity are to (1) analyze knowledge, values and motivations about sustainability challenges in the watershed and (2) explore how they relate to priorities for action in context of a diverse community of stakeholders with competing values and priorities.
2. Describe sustainability challenges (5–10 min). Explain that the class is going to use a set of statements to understand the complexities of sustainability in the context of managing a watershed and water use. Hand out the statements.
3. The Q-sort process (3–5 min). Explain that each student is going to sort the statements to create a personal ranking using a method that is used in real-world sustainability research and management. Explain that Q-sort is a nice compliment to other survey methods because it captures the ways that people think about ideas in relation to one another rather than in isolation. Q-sort and other similar social science methods are key to understanding sustainability challenges because issues can be so intricately linked. A limitation of Q-sort is that the results do not reveal the prevalence of different ways of thinking in the population at large.
4. Rank statements (20–30 min). Students will read each statement. If applicable, the student will cut out each statement as they read. Then, each student will create three initial piles based on whether they think each statement should be a high sustainability priority or not, or are unsure. Students will record the number of statements in each pile and then further sort statements using the sorting pyramid. Students fill in the sorting pyramid beginning at the edges and alternating between

the high and low priority sides until they reach the middle. The presorted piles speed the task and the "unsure" pile can be used to fill in any gaps. After they have finished and settled on their rankings, students record the number (1–30) printed on each statement in the box in the sorting pyramid that corresponds to its ranking (A-AD). Students then answer the open-ended questions (ESM-D) which are designed to elicit their values, motivations, knowledge, and other factors that may have influenced the statements' rankings. The rationale for these choices demonstrates how each student defines and operationalizes sustainability.

5. Compare and contrast results (10–20 min). This is intended to replicate the principal components analysis researchers would use to group similar responses. With a small class that meets in a classroom that is easy to move around in, ask students to move around the room and, using their sorting pyramids, create groups based on the similarities of their ranking results. Students should interact with as many peers as possible before establishing groups so that they are aware of the range of approaches to ranking. Instructors may choose to tell students that they are, in essence, performing a very informal version of the principal components analysis that would normally be used to analyze the pyramids. Full instructions and an alternative option for larger courses are provided in ESM-A.

6. Reconvene and discuss (10–20 min). Ask students to share the how they formed and defined their groups. The goal of the discussion is to highlight the differences in knowledge (what, why, and how) that differentiated groups. Even without substantial background preparation, students are likely to have life experiences that led them to make inferences about the relative priority of statements for which they had little information. These values, beliefs, and experiences provide context and legitimacy to information and facts. The instructor may choose to discuss how Meo et al. (2002) used Q-sort as an example of its application to other water-management scenarios.

7. End the activity (5 min). To close, summarize the main learning objectives of the activities and details about the uses and limitations of Q-sort. Subjective points of view matter because they shape how people interpret "the facts." Therefore, negotiation among people with different values may include sharing information about "the facts" but it also requires understanding what is important to others and why so that collectively, stakeholders can find creative ways to limit trade-offs. A list of other key points is provided in the ESM-A. To close, ask students to produce a written reflection, explaining how a group might proceed when visions of sustainability are not shared (or some stakeholders reject the idea of sustainability entirely). This is question 6 in ESM-D. Discuss the responses.

Follow-Up Engagement

- To extend and enhance their learning, students can be asked to:
  - Repeat the Q-sort procedure while acting as researchers who are collecting, evaluating and reflecting on responses from peers outside of class.

- Read a journal article that applies Q-sort and write a reflection on the findings of the paper based on their in-class experience as a "research subject" (e.g., Meo et al. 2002; Previte et al. 2007; Webler et al. 2009).
- Discuss case studies in environmental management that reveal how differences in worldviews were important to effective action (e.g., Webler et al. 2009).

- Instructors could also use the data generated in the activity to teach qualitative and quantitative data analysis techniques such as content analysis, descriptive statistics, and principal components analysis (e.g., Pruslow and Red Owl 2012).
- Other chapters in this volume provide related activities to help students explore worldviews (Chaps. 4 and 39), management of water-related resources (Chaps. 13 and 16) and land use planning and management (Chaps. 17 and 20).

Connections

This exercise relates to the examination of:

- Environmental policies such as the US National Environmental Policy Act (NEPA) that call for public engagement in environmental decisions.
- Stakeholder engagement and community-based management of common-pool natural resources (e.g., Chap. 16).
- The relationship between worldviews and values for determining sustainability outcomes (e.g., Chaps. 4 and 39).

References

(USGS) United State Geological Survey (2014) Upper Midwest Environmental Sciences Center: About the Upper Mississippi River System. http://www.umesc.usgs.gov/umesc_about/about_umrs.html. Accessed 7 Jul 2015

Danielson S, Webler T, Tuler SP (2009) Using Q method for the formative evaluation of public participation processes. Soc Nat Resour 23:92–96. doi:10.1080/08941920802438626

Delong MD (2005) Upper Mississippi River basin. In: Benke AC, Cushing CE (eds) Rivers of North America. Elsevier Academic Press, Burlington, MA, pp 327–373

Meo M, Focht W, Caneday L et al (2002) Negotiation science and values with the stakeholders in the Illinois River Basin. J Am Water Resour Assoc 38:541–554

Previte J, Pini B, Haslam-Mckenzie F (2007) Q methodology and rural research. Sociol Ruralis 47:135–147. doi:10.1111/j.1467-9523.2007.00433.x

Pruslow JT, Red Owl RH (2012) Demonstrating the application of Q methodology for fieldwork reporting in experiential education. J Exp Educ 35:375–392. doi:10.5193/JEE35.2.375

Webler T, Danielson S, Tuler S (2009) Using Q method to reveal social perspectives in environmental research. Social and Environmental Research Institute, Greenfield, MA, www.seri-us.org/pubs/Qprimer.pdf. Accessed 7 Jul 2015

Wiek A, Withycombe L, Redman CL (2011) Key competencies in sustainability:. a reference framework for academic program development. Sustain Sci 6:203–218. doi:10.1007/s11625-011-0132-6

Chapter 20
Location, Location, Location! Analyzing Residential Development in Environmentally-Fragile Areas

Erica A. Morin

Introduction

After World War II, the USA experienced a massive housing shortage due to the influx of returning soldiers, increasing incomes, and "booming" family sizes. The resulting need for homes, coupled with a lack of local, state, and federal land-use regulations, led to widespread residential construction in environmentally-fragile areas including wetlands, hillsides, and floodplains. These construction practices caused significant groundwater pollution, destruction of wildlife habitat, and soil erosion in many communities (Rome 2001; Kahn 2000). Eventually, some towns, counties, and states mandated construction permits and zoning standards, but they were not exhaustive, and economic concerns of development frequently trump environmental ones to this day (Pendall 1999; Samuels 2015). Despite clear environmental concerns and risk of natural disasters, residential construction continues unabated in many environmentally-fragile areas due to desirable urban proximity, community status, and amenities like vistas and waterfronts (Van Der Vink et al. 1998).

For the activity described below, students engage in an inquiry-based group exercise in which they use personal electronic devices to research and assess the risks of residential development in preselected environmentally-fragile areas. The goal is for students to gain a better understanding of the environmental considerations of residential development siting, planning, and construction. Through further discussion and follow-up assignments, the learned insights can help students draw conclusions about the importance of education and regulations to limit the extent and style of development in certain ecosystems and topographic circumstances.

Electronic supplementary materials: The online version of this chapter (doi:10.1007/978-3-319-28543-6_20) contains supplementary material, which is available to authorized users.

E.A. Morin (✉)
Department of History, Westfield State University, Westfield, MA, USA
e-mail: emorin@westfield.ma.edu

Learning Outcomes

After completing this activity, students should be able to:

- Discuss the environmental implications of siting, planning, and building a house or residential subdivision/neighborhood.
- Identify environmental characteristics that make certain locations ill-suited for residential development.
- Hypothesize about how to encourage and/or require better land-use planning and residential construction practices in environmentally-fragile areas.

Course Context

- Developed for an undergraduate US environmental history course with 30 students, minimum sophomore standing, no course prerequisites, and a wide range of majors
- 50 to 75 min in one class meeting
- Students should have a basic knowledge of ecological conditions affecting geographic regions throughout the USA that can be provided via preparatory lectures or reading assignments
- This activity could be modified for use in a larger class with the help of teaching assistants to guide students

Instructor Preparation and Materials

When used for the first time, this activity requires substantial instructor preparation, but it can subsequently be repeated with little adjustment. Instructors should begin by familiarizing themselves with the process of home construction and certain environmental areas that are extremely fragile and ill-suited for residential development (see Electronic Supplementary Materials (ESM) A). For example, it is useful to have a basic knowledge of the steps of siting and building a home, particularly related to the property's soil composition, slope, and proximity to water, wetlands, and other natural resources. Additionally, the instructor should incorporate the risk of natural disasters, such as floods, droughts, tornados, hurricanes, earthquakes, and wildfires into the activity development. It is also recommended that instructors preview a brief video about Modern Home Construction Techniques, which is also useful to show in class during the discussion (Civil Engineering 2012).

The instructor should choose and conduct preliminary research about 10–15 specific communities (one for each group of two to four students) around the country that are situated in environmentally-fragile areas and/or had experienced severe natural disasters in the recent past (e.g., Table 20.1 and ESM-B). These

Table 20.1 Examples of towns in environmentally-fragile areas

• Beaufort, NC (hurricanes, barrier islands, coastal marshes)
• Cairo, IL (flood plains)
• Cascade, CO (wildfires)
• Greensburg, KS (tornados)
• Hollister, CA (earthquakes)
• Homestead, FL (wetlands, plant and wildlife diversity)
• LaConchita, CA (slopes, landslides)
• Mead, CO (fracking)
• Mears, MI (sand dunes)
• Yerington, NV (groundwater)

towns should have easily identifiable environmental issues that students can readily find through online searches. Select a range of locations prone to landslides, floods, wildfires, earthquakes, droughts, tornadoes, and hurricanes, and places near swamps, marshes, coastlines, sand dunes, wildlife migration zones, and similar ecological concerns. The instructor should know the characteristics of these locations in order to help guide the student research activity. During class planning, the instructor should put him/herself in the position of the student and conduct a basic online search of each potential town to determine the contents of the first page of search results, including the images, recent news stories, Wikipedia entry, and town homepage, if one exists. In the author's experience, the initial search results should display the location, natural surroundings, recent environmental issues, and occurrence of natural disasters with relative ease. The students should not be expected to have advanced research skills, so the first page of search results should provide them with enough information to draw basic conclusions. (For example, an online news search for "Cascade, Colorado" specifically mentions the Waldo Canyon Wildfire of 2012, which was the environmental risk for students to identify in this town.) The list of communities can be easily updated each semester, depending on the focus of the course and to reflect recent environmental topics and disasters. The supplementary presentation found in ESM-C corresponds directly with the communities in Table 20.1 and ESM-B, but can be edited to reflect instructor-chosen additions and omissions.

To fully implement the activity, the classroom should have digital projector capabilities (for presentation slides), a chalk/dry erase board (for recording brainstorming notes) and wireless Internet connection to enable students' online research. Students must have at least one portable, Internet-accessible device per group (smartphone, tablet, or laptop) to complete the research. The instructor can poll students prior to planning the activity to assess the availability of devices among students, and remind them to bring these devices on the day of the activity. (If the availability of Internet-capable devices is a concern, the instructor should prearrange the groups to ensure even distribution of devices.) If wireless connectivity is unavailable or there is a shortage of devices, the activity could alternatively be completed in a computer lab. The activity also requires students to complete a worksheet, one per group of two to four students (see ESM-D). Instructors should review the worksheet and make necessary edits before printing or photocopying.

Before the activity, students should have a basic knowledge of the geography and climate of the various regions of the country or area to be discussed. In the author's course, students had preexisting knowledge of the geographic regions of the USA and their ecological conditions from previous lectures and map quizzes. If time does not allow for a separate lesson about these topics, the instructor may wish to assign a preparatory reading to help familiarize students with the regions and climates. The Environmental Protection Agency's Ecoregions maps and Natural Disaster information pages are useful resources for this task (see ESM-A). Additionally, depending on class period length and focus of the course, the instructor may wish to assign a reading in advance to prepare students for the activity (see ESM-A for suggestions), and allocate time (~10–15 min) for explicit discussion of it between or within the steps described below.

Activities

The full lesson is implemented in three parts, as follows. Overall, the activity begins with a full-class discussion of the steps of residential development and corresponding environmental concerns, then students shift into small groups to research specific towns, and lastly, students regroup to share findings with the entire class.

1. Brainstorming (~15–20 min) about home construction and broader environmental implications

Begin with this activator question to stimulate students' thought processes and establish context for the research: what are the steps involved in building a house? This activator is especially useful for introductory courses with a wide range of majors because it helps establish a common starting point. This question can be written on the board or displayed on a slide (see slide 2 in ESM-C). Ask students to write a response to this question based on their current knowledge. After this, engage the class in an informal group discussion about how a house is built. Instructors should have the students share ideas about the steps and compile a roughly sequential list on the board (see diagram of brainstorming activity in ESM-E). Important steps include: site selection and preparation, architectural style, excavating the foundation, framing and construction materials, installing utilities, heating and cooling needs, and landscaping. Next, if desired, play the brief video that shows the process of modern home construction techniques from start to finish (embedded in slide 3). This discussion and video helps students visualize how residential construction affects and is affected by the terrain, soil, bedrock, water, temperature, weather, and other environmental factors. For example, many students do not consider the issues associated with digging into the ground, building on an inclined surface, or how to handle the presence of nearby water, among other concerns.

After the introduction, the instructor should transition to discussing the environmental implications. While the initial activator question allows students to grasp the process of home building at the individual level, further discussion expands the topic and students' perspectives by connecting the effects of building one home to the environmental consequences of construction processes in various ecological, climatological, and geographic areas. Using the previously generated list of steps, ask the class to think about what environmental factors need to be considered in connection with each step. The instructor can write corresponding environmental concerns in a second column next to the construction steps and/or draw arrows to connect steps and issues (see ESM-E). For example, creating a house's foundation depends on soil, bedrock, slope, and the presence of surface and ground water. The instructor may need to ask leading questions to direct student comments. Using images of homes from the presentation (see slides 4–7) can also help generate student comments about the environmental factors associated with construction.

Next, encourage students to think about the cumulative impact of residential development. While building one home may not seem significant, building hundreds or thousands of homes in a small area carries greater environmental burdens. Due to lower land costs, many residential development projects are built on environmentally-fragile land and thereby increase the negative impact on the surrounding ecosystem. Most students recognize that the environmental risks increase as the number of home constructed increases. Utilize the presentation (slides 8–26) to help students think about a variety or geographic and ecological circumstances that affect and are affected by large scale residential development, including farmlands, wetlands, hillsides, and floodplains. Most students envision the types of homes that are most common in their geographic area, but the instructor can encourage them to picture homes in different regions. While homes in New England often have basements and cement foundation, homes in more humid and sub-tropical areas (e.g., Louisiana, Florida) are built above ground or on stilts because of excessive water concerns. In areas that experience certain types of natural disasters, such as "tornado alley" in the American Midwest, a basement is beneficial for safety reasons while certain house types, like mobile homes, have higher risks of damage. Most students can draw on common knowledge and common sense to brainstorm these steps and considerations, but a pre-class reading assignment can help them prepare for generating the list, as desired (see ESM-A for suggestions).

2. Group research (~15–20 min)

After creating a list of the environmental effects of home construction and the compounded risks of large scale residential development in a variety of environmental areas, the students will engage in a research activity. The instructor should place the students into small groups (of two to four, randomly or preassigned) to assess the risks of residential development in specific locations.

Each group is given one town name and a worksheet (ESM-D) to complete. One group member may be designated as recorder but all students' names should be included on the worksheet.

Ask students to imagine that they are a team of residential contractors and developers that have been hired to determine the environmental conditions and risks of construction in their specific community. The instructor should explain that each community is located in an environmentally-fragile area and/or regularly experiences a specific environmental issue or natural disaster risk. Student groups are instructed to look up the town on their devices and complete the worksheet about the area's characteristics. Students should be informed that not all the questions are directly applicable to their town, but each location does have a serious environmental concern. Also, the instructor should inform students that they will be asked to share their group's findings at the end of the class period.

Following the organization of the worksheet, students should begin by determining the town's location which can provide a great deal of information about geography and climate. It is acceptable and assumed that students will use the town's Wikipedia page and other basic sources to derive this information. To find more details, students should then search using keywords such as the "town name, state" coupled with "environment," "ecosystem," "natural disasters," "environmental problems," or more specific key terms that students glean from preliminary findings, such as "hurricane," "drought," "wildfires," "tornados," etc.

Assuming that the instructor has done pre-class research on the towns, she should be able to circulate throughout the classroom to direct students toward relevant information as needed and help with their worksheet responses. In addition, because this activity can present an enormous temptation for students to use their devices for reasons other than research, the instructor is encouraged to monitor students to help them stay on task and minimize distractions. In the author's experience, for the most part students have generally been excited to use their devices in class and were respectful of the expectations.

3. All-class discussion to summarize findings (~15–20 min)

After the groups have completed all or most of their work, the class should be asked to regroup to share their findings. Students are asked why construction is not desirable in each location. As they respond in turn, the instructor can display slides with maps of each town's location and pictures of its natural setting (see slides 28–48 in ESM-C for map and images of the towns in Table 20.1). After reviewing the diverse case studies, students should better understand how residential development in environmentally-fragile areas poses risks to ecosystems and residents in every part of the country. Residential development in environmentally-fragile areas can be seen, therefore, as an epidemic in the USA, not simply a local zoning issue.

After reviewing several locations with poor building prospects, the instructor may ask (if students have not already): why do people keep building houses in these places? In the author's classes, students' answers have reflected scholarly research findings, including proximity to nice areas, desirable vistas, and seeking community social status. An additional key question is: how can communities/towns prevent this type of environmentally destructive development? This question is completely open-ended and can be incorporated into future lectures, discussions, or assignments, as desired.

Follow-Up Engagement

- These questions can be used for additional discussion or follow-up assignments:
  - What new information did you learn about the environmental risks of housing development and suburban sprawl through this activity?
  - What does continued construction in environmentally-fragile areas suggest about societal valuation of the natural world?
  - Should homes in environmentally-fragile areas be rebuilt after natural disasters?
  - What kinds of laws/policies/incentives could be created to limit this type of development?
  - Do you think this information will influence your decisions about where to live in the future?
- A writing assignment can be developed based on the nature and content of the class discussion. For example, students could be asked to replicate the skills learned in the activity on their own. Students could be asked to find the worst location for residential development in the USA or be challenged to propose a residential development location that poses little to no environmental risks.
- Activities are described in this volume that can be used to explore additional issues relevant to urban planning including watershed management (Chap. 19), place-based stewardship (Chap. 17), environmental justice (Chap. 30), pollution (Chaps. 31–33), and green roofs (Chap. 34).

Connections

This activity relates to many other environmental and sustainability topics in other disciplines, including:

- History/Political Science: the growth of environmental legislation and regulations in the 1970s.
- Natural Resource Management/Environmental Science: the impact of residential development upon water resources, wetlands, wildlife habitat, and natural disaster relief (for related activities see Chaps. 13, 17, 19 and 34).
- Engineering/Building Construction Management: improving construction methods to minimize or avoid landslides, groundwater contamination, excessive energy consumption, and negative environmental impacts.

- Geography/Planning: land-use planning and management strategies to curb the environmental damage caused by increased residential development and restrict building in certain areas.

References

Civil Engineering (2012) Modern home construction technique. www.youtube.com/watch?v=FQvFzdFIp08. Accessed 10 July 2015

Kahn ME (2000) The environmental impacts of suburbanization. J Pol Anal Manag 19:569–586

Pendall R (1999) Do land-use controls cause sprawl? Environ Plann B Plann Des 26:555–571

Rome A (2001) The bulldozer and the countryside: suburban sprawl and the rise of American environmentalism. Cambridge University Press, New York

Samuels A (2015) Why are developers still building sprawl? The Atlantic, February 24. http://www.theatlantic.com/business/archive/2015/02/why-are-people-still-building-prawl/385741/. Accessed 10 Apr 2015

Van Der Vink G, Allen RM, Chapin J et al (1998) Why the United States is becoming more vulnerable to natural disasters. Eos Trans AGU 79:533–537. doi:10.1029/98EO00390

Part III
Consumption, Economics and Energy

Chapter 21
Tasting Sustainability: Using Multisensory Activities to Retune Food Preferences

Jennifer Gaddis

Introduction

Processed foods now account for 70 % of the calories consumed by Americans (Warner 2013). Powerful multinational companies profit from consumer reliance on this industrialized food system. The cheaper consumer price of this diet—as opposed to local, organic, made-from-scratch alternatives—does not provide a true accounting of the ecological or social costs of "cheap" food (Carolan 2011, Food Economics n.d., Doe 2015).

High rates of soil loss, groundwater withdrawal, and burning of fossil fuels point to the ecological unsustainability of industrial agriculture (Heller and Keoleian 2003). Likewise, sociological trends like ballooning health costs, the decline of family farms, the "graying" of American farm owners, increasing reliance on migrant laborers, and high rates of economic insecurity among food chain workers all point to the social unsustainability of the current system. At the planetary scale, the industrial food system threatens the biophysical conditions necessary for human life; climate, biodiversity, and the nitrogen cycle are all already outside of the safe operating space for humanity in part due to transformations of the biosphere caused by agricultural intensification (Rockström et al. 2009).

Electronic supplementary materials: The online version of this chapter (doi:10.1007/978-3-319-28543-6_21) contains supplementary material, which is available to authorized users.

J. Gaddis (✉)
Department of Civil Society and Community Studies, University of Wisconsin-Madison, Madison, WI, USA
e-mail: jgaddis@wisc.edu

© Springer International Publishing Switzerland 2016   165
L.B. Byrne (ed.), *Learner-Centered Teaching Activities for Environmental and Sustainability Studies*, DOI 10.1007/978-3-319-28543-6_21

A sustainable food movement has already begun to emerge, as evidenced by the growth of low-carbon diets, farmers' markets, community-supported agriculture, networks of young and beginning farmers, fair trade and organic certifications, worker-led campaigns for economic justice, and animal welfare initiatives. But for sustainable food to move from niche to mainstream, public policy needs to lower the geographical and economic barriers to purchasing these foods. In addition, individual and societal food preferences must shift away from the "flattened out" tastes and expectations of industrial food (Carolan 2015), which pose an "affective barrier" to the adoption of sustainable food practices. For behaviors to change, alternatives must also be *felt*, making active "retuning" of everyday tastes and expectations critical to the transition process (Carolan 2015).

The activity in this chapter uses a guided food tasting to prompt students to think about retuning their own taste preferences while stimulating a broader discussion about affective barriers to sustainability (also see Chap. 6, this volume). During the tasting, students will mindfully observe sensations of sight, smell, sound, taste, and touch. This provides a low-stakes opportunity for students to experiment with moving beyond "knowing" to actually "feeling" sustainability. In general, multisensory-learning activities offer a promising strategy for sustainability studies, given that such activities increase students' engagement, foster deeper participation, and model learning as a fun experience (Baines 2008).

Learning Outcomes

After completing this activity, students should be able to:

- Reflect on their personal taste preferences and get a "feel" for sustainable food using the practice of mindful eating to focus their attention.
- Recognize "affective barriers" that make sustainable food choices more difficult or unlikely.
- Explain why retuning taste preferences matters for the sustainability of food systems.

Course Context

- Developed for an introductory food and sustainability course for community studies majors and a class size of 20–30 students
- With an introductory discussion of 15 min, this activity can be completed in one 50-min class period and can be extended with a longer discussion of the assigned reading if desired
- Before class students should read Carolan (2015), but no additional background knowledge is needed

- Adaptable to courses of any class size and level that include discussions of food consumption or sustainable lifestyles; the "mindful tasting" exercise can be modified for large-class sizes by limiting the number of samples and choosing a food that is easy to distribute to large groups

Instructor Preparation and Materials

To begin, the instructor should decide how much time to allocate to in-class discussion of the recommended assigned reading (Carolan 2015) and the tasting. (The text below describes the implementation for one stand-alone tasting, but instructors who wish to conduct multiple tastings or to pair tastings with related readings should consult Electronic Supplementary Materials (ESM-A and ESM-D.)

Prior to the activity, instructors should read Carolan's (2015) article, which argues that people need to be "retuned" to the tastes, textures, cares, and practices associated with alternatives to the dominant industrial food system. Instructors who wish to gain a deeper understanding of the alternative food movement can read Follett (2009), Goodman and Goodman (2009), Holt-Giménez and Wang (2011), and Lockie (2009). For those looking to learn more about the social determinants of consumer food choices, see Jaffe and Gertler (2006) and Vermeir and Verbeke (2006) for a discussion of the gap between consumers' attitudes, behaviors, and intentions to consume sustainable foods. Finally, to learn how visceral experience with sustainable food can mobilize people to participate in social and environmental activism, see Hayes-Conroy and Martin (2010). Other background readings can be found in ESM-A.

For the tasting exercise, the instructor should identify a type of food or drink and then procure two to five different samples that will allow students to distinguish between the "flattened" taste of industrial foods and that of alternative production methods. Attributes to look for when selecting samples for the "sustainable" food category include artisanal, local, organic, fair trade, non-GMO certified, all natural ingredients, seasonal, and heirloom/heritage (i.e., plant varieties or animal breeds that differ from the most common, standard varieties). For example, a simple tasting of two samples might feature a store-bought, container-ripened, Red Delicious apple vs. a locally grown, freshly picked heirloom apple from a farmer's market. Sufficient amounts of the chosen food should be obtained for each student in the class to have a tasting portion. Thus, another key factor to consider is class size; foods that require special utensils or refrigeration may be difficult to use in large classes, whereas food choices like apple slices, radishes, and crackers are easy to use in classes of any size.

The instructor will also need to review and, as desired, edit or prepare alternatives to the suggested in-class discussion questions (in SEM-B) for the Carolan (2015) reading which should be assigned to students before the tasting class. (These questions can be printed on a handout for students to complete in class.) In addition, prepare and print a tasting guide (one for each student) that includes vocabulary to help students describe what they see, feel, smell, and taste. The guide should also provide a space to write tasting notes about each sample (see ESM-C for a butter tasting example). Finally, all food samples (and any needed utensils) must be prepared for distribution to students and clearly labeled (with a name or number if the instructor wishes to conduct a blind tasting).

Activities

1. Retuning "flattened" tastes (15–50 min, depending on the length of the article discussion)

   To begin the class, lead a discussion about Carolan (2015) (e.g., using questions from SEM-B). The instructor can pose the questions to the whole class or ask students to answer them in small groups, possibly writing responses on a handout.

2. Mindful tasting (5–20 min, depending on the number of samples)

   Explain to students that they are invited to participate in a guided "mindful tasting," which will provide them with an opportunity to experience and reflect on the type of "retuning" that Carolan (2015) advocates as a mechanism for overcoming affective barriers to sustainability. Pass out the samples (as needed, asking for student volunteers) and tasting sheets. Remind everyone that ingesting the samples is purely voluntary. (Students with allergies or food aversions should be instructed to participate in the activity using their senses of sight, touch, and smell if they are able.)

   Throughout the tasting, instruct students to do their best to isolate and name the sensations they are experiencing using the vocabulary on the handout. Assure them that there are no wrong answers, and direct them to record their honest reactions on the tasting sheets.

   Lead the students through the steps of mindfully consuming each sample using these prompts:

   (a) First, look at the sample and notice its color and texture.
   (b) Next, close your eyes and touch the sample. Pay attention to how it feels. Is it hard or soft? Smooth or rough? Moist or dry?
   (c) Now, smell the sample. What scents can you detect?
   (d) Finally, consider its sound and taste. Make sure not to eat it all in one bite, since you might want to return to the sample to do a comparison after you've tasted the others. Take a bite and chew very slowly, noticing all of the sensory experiences of chewing and tasting. What does it sound like when you bite into it and begin to chew? How does it feel in your mouth? How does it taste? Try to isolate some particular texture and flavor notes. Does the intensity of the flavor or the texture change the longer it's in your mouth?

Repeat the tasting prompts for each sample. After the last one, ask students to go back through and taste the samples again at their own pace, this time specifically focusing on samples they'd like to compare. Encourage them to share their reactions with their neighbors. When students have tasted and compared all the samples, ask them to identify their favorite and least favorite. If the tasting was blind, ask them to guess which sample is which (from a list of choices on the board), then reveal the correct answers. If the students seem surprised by their own preferences, make note of this, and in the debrief discussion, use this to spark a discussion about what they think they "should" prefer and how or why their palates may have tricked them.

3. Debrief (15–25 min, depending on how long students are given to work in small groups)

After the tasting, lead a short discussion to debrief the activity. Ask students to share their reactions to the different samples. Help draw out strong reactions and pay attention to the presence of "ambivalence," reminding students that tastings don't necessarily elicit an immediate change in perception or behavior, but are nonetheless an important tool for generating new questions and potentially destabilizing established attitudes and expectations. If time allows, instructors might consider referencing key passages about ambivalence from the Carolan (2015) text (pp. 326 and 327).

Next, ask students to pair up (or form small groups). Instruct them to quickly name a few barriers to sustainable food consumption that operate "out there" in society and affective barriers that operate "within" a person. Then solicit answers and write them on the board. As a whole class, categorize the responses as they relate to factors such as cost, physical access, shopping and preparation time, marketing, nutrition, ecological sustainability, social justice, taste, and personal experiences.

Finally, end with a discussion of how to help more people "get a feel for the alternatives", as Carolan (2015) advocates. Ask the students to return to their pairs/small groups and to brainstorm a list of policies or programs that would create the conditions for more people to participate in alternative agriculture and food systems. If desired, have each group or student write a bullet-point summary of the recommendations.

Follow-Up Engagement

- Compile the student-generated list of policies and programs that will help people "get a feel for the alternatives." Identify the most frequently proposed ideas and lead a discussion about what it would take to actually implement the ideas on their campus/in their community.

- Conduct an in-class simulation to see if students are willing to pay more for products that are marketed as sustainable (see Vecchio and Annunziata 2015 for guidance; also see Chaps. 22 and 23, this volume).
- Ask students to conduct a tasting with a group of friends. Have them write a brief reflection on their experience, including a description of the participants' reactions and whether they think tastings are a promising way to lower affective barriers to sustainability.
- Pair the tastings with local field trips. In the original course, sausage and coffee tastings were paired with field trips to a local butcher shop and coffee roaster that exemplify best practices in ecological, economic, and social sustainability.
- Have students give an oral report or write a brief reflection on a food of their choosing. See ESM-D for sample prompts.
- Lead the "What's in your freezer" activity (ESM-E), which fosters interesting discussion about food choices and helps students connect with each other on a personal level.

Connections

- The tasting activity introduces students to the practice of mindfulness, given that they are asked to focus their awareness on the present moment and to acknowledge and name their own feelings, thoughts, and bodily sensations. The relationship between mindfulness and sustainability, across a wide range of practices and topics, is a ripe area for experiential education and classroom discussion about lifestyle change. When people slow down and intentionally observe, what do they notice differently? How does focusing attention inward, and allowing oneself time to notice sensation, foster a calm mental state and contribute to general well-being? According to Ericson et al. (2014), promoting mindfulness increases subjective well-being, empathy, and an examination of personal values, all of which can lead to more sustainable behaviors and to greater human well-being (see Chaps. 4, 6, 8, 29 and 43, this volume, for related topics and activities).

References

Baines L (2008) A teacher's guide to multisensory learning: Improving literacy by engaging the senses. ASCD
Carolan M (2015) Affective sustainable landscapes and care ecologies: getting a real feel for alternative food communities. Sustain Sci 10:317–329
Carolan M (2011) The Real Cost of Cheap Food. London: Earthscan.
Doe J (2015) Food economics. In: Grace communications foundation. Food Program. http://www.sustainabletable.org/491/food-economics. Accessed 1 June 2015
Ericson T, Kjønstad BG et al (2014) Mindfulness and sustainability. Ecol Econ 104:73–79

Follett JR (2009) Choosing a food future: differentiating among alternative food options. J Ag Environ Ethics 22:31–51

Goodman D, Goodman M (2009) Alternative food networks. In: International encyclopedia of human geography. Elsevier, Oxford, pp 208–220

Hayes-Conroy A, Martin DG (2010) Mobilizing bodies: visceral identification in the slow food movement. Trans Inst Br Geogr 35:269–281

Heller MC, Keoleian GA (2003) Assessing the sustainability of the US food system: a life cycle perspective. Ag Systems 76:1007–1041

Holt-Giménez E, Wang Y (2011) Reform or transformation?: the pivotal role of food justice in the US food movement. Race/Ethnicity Multidisciplinary Global Contexts 5:83–102

Jaffe J, Gertler M (2006) Victual vicissitudes: consumer deskilling and the (gendered) transformation of food systems. Ag Human Val 23:143–162

Lockie S (2009) Responsibility and agency within alternative food networks: assembling the "citizen consumer". Ag Human Val 26(3):193–201

Rockström J, Steffen W, Noone K et al (2009) A safe operating space for humanity. Nature 461:472–475

Vecchio R, Annunziata A (2015) Willingness-to-pay for sustainability-labeled chocolate: an experimental auction approach. J Cleaner Prod 86:335–342

Vermeir I, Verbeke W (2006) Sustainable food consumption: exploring the consumer "attitude–behavioral intention" gap. J Ag Environ Ethics 19:69–194

Warner M (2013) Pandora's lunchbox: how processed food took over the American meal. Simon and Schuster, New York

Chapter 22
Relationships Between Consumption and Sustainability: Assessing the Effect of Life Cycle Costs on Market Price

Madhavi Venkatesan

Introduction

Economics is a science that evaluates human behaviors within the framework of a resource-constrained world to assess the use of, limits to, and, thereby, sustainability of resources (for additional discussion, see Daly 1996, 2014; Dietz and O'Neill 2013; Greer 2011). However, because many behaviors (i.e., immediate gratification, conspicuous consumption) can be characterized as legacies, inherited from previous generations, they may become outdated and promote undesirable human activities that interfere with the long-term success of economic systems and ultimately the sustainability of human and nonhuman life (for additional discussion, see Electronic Supplementary Materials (ESM)-A). As a result, unless there is continuous assessment of the rationale for individual and societal behaviors, needed changes can be delayed, leading to or maintaining the persistence of unsustainable outcomes, as defined by resource overexploitation, degradation, and depletion (Hards 2011; Venkatesan 2015). From this perspective, given today's world of increasing human population size and consumption rates, a need exists for increased awareness of the motivations for present behaviors and an evaluation of the same relative to the promotion and achievement of sustainable outcomes (e.g., see Chaps. 4, 6, and 21, this volume).

The goal of the assignment described in this chapter is for students to analyze their individual consumption motivations (i.e., needs, wants, marketed demands, etc.) and evaluate the sustainability of a specific consumption choice. The assignment is a qualitative assessment of the life cycle costs of a purchased

Electronic supplementary materials: The online version of this chapter (doi:10.1007/978-3-319-28543-6_22) contains supplementary material, which is available to authorized users.

M. Venkatesan (✉)
Department of Economics, Bridgewater State University, Bridgewater, MA, USA
e-mail: madhavi.venkatesan@bridgew.edu

© Springer International Publishing Switzerland 2016
L.B. Byrne (ed.), *Learner-Centered Teaching Activities for Environmental and Sustainability Studies*, DOI 10.1007/978-3-319-28543-6_22

good that, by definition, evaluates environmental and social costs (i.e., externalities) from production to disposal (also see Chap. 2, this volume). Students conduct online research about the life cycle impacts of a good to compare the externalized costs with the good's observed market price. In the assessment process, students are prompted to determine whether the price promotes overconsumption and, via externalities, exacerbates resource depletion, degradation, and exploitation. Through the assignment, student awareness of the significance of consumerism along with manufactured wants, stemming from marketing, advertising, and the media or cultural values that promote consumption as a leisure activity, is increased through class discussion. With this assignment, instructors can increase students' tangible awareness of the inherent responsibilities embedded in consumption behavior. The pedagogy used promotes research, collaboration, self-evaluation, and appreciation of the nexus between economics and other disciplines as a critical dimension of the policy and assessment mechanism requisite in the assessment of environmental and sustainability issues.

Learning Outcomes

After completing this assignment, students should be able to:

- Recognize the current drivers (e.g., marketed demand, supply, want, need, price) of their own and others' consumption behaviors.
- Assess the market price of a good relative to the externalities of its life cycle (production, distribution, consumption, and waste).
- Describe how market prices impact (un)sustainable consumption behaviors.

Course Context

- Developed for introductory economics course with 25–35 students, primarily noneconomics majors
- Two 50-min class meetings separated by sufficient time for students to engage in outside-of-class research, but adaptable to a single class meeting
- No background knowledge is needed prior to introduction of assignment

Instructor Preparation and Materials

The following text describes the full implementation of the assignment. As desired, it can be modified to fit within one class period, for example, by using videos to introduce concepts outside of class (e.g., those in ESM-B) or assigning writing assignments instead of class presentations (see "Follow-Up Engagement" section).

To facilitate the introductory background for the assignment, the instructor should prepare to provide students with a foundation of the economics concepts provided in Box 22.1 and the three assumptions embedded in the classical definitions of demand and supply: (1) "unlimited wants" on the part of the consumer, (2) market price efficiency, which is the implied view that market price at equilibrium adequately assesses production and consumption costs, and (3) producer motivation to minimize costs and maximize revenue (profit), in part by maximizing externalities, often with unsustainable outcomes. These concepts are explained in more detail in ESM-A alongside an example explanatory narrative that can be used to present them to students. A key issue to convey to students is how externalities relate to the price of a good or service, whereby non-quantified costs (i.e., externalities) essentially subsidize (i.e., reduce) the market price and thereby promote unsustainable resource utilization rates (for additional discussion and perspectives, see Hards 2011; Fischer et al. 2012; Sen 2010; Venkatesan 2015; for a sustainability-focused primer on these concepts, see Venkatesan 2016). Additionally, the instructor may choose to review the online videos provided in ESM-B and to share the same with students to facilitate understanding of how the topics of consumerism, economic growth, externalization of costs, and GDP relate to sustainability. The introduction of foundational economics concepts provides an opportunity to encourage general class discussion related to students' personal assessment of what motivates their consumption choices (e.g., needs, wants, marketed demands, etc.).

Box 22.1: Foundational Economics Concepts (See Further Discussion in ESM-B)

- Definition of economics
- Definitions of supply and demand, law of supply, law of demand and embedded assumptions in supply, and demand curves
- Historic evolution of gross domestic product (GDP) from production capacity measure to an indicator of global economic progress
- Definition of the expenditure equation of GDP
- Assumptions of consumer and firm (i.e., business or producer) behavior
- Assumption of equilibrium market price related to price accuracy and the cost of products and services
- Potential for market prices to under-remunerate true costs and thereby allow for unsustainable consumption
- Definition of externality and how externalizing of costs facilitates unsustainable consumption
- Life cycle analysis

As this activity involves student assessment of the life cycle of a consumption good, the instructor should determine what good(s) would provide the best outcome for the course keeping in mind that the good(s) should be an item that all students would have a *willingness and ability* (which are defining concepts embedded in assumptions of demand) to purchase. (Alternatively, students could be allowed to choose their own goods to investigate.) In addition, to facilitate the discussion of externalities and sustainability, the good should have potential production costs not included in the price (e.g., carbon and water footprints, effects on human health, and natural resources, all of which are listed in the assessment table as provided in ESM-C to be given to students) that will reflect underpricing of the good, thereby promoting higher consumption rates and faster depletion of resources. In the author's classes, an individually consumable beverage (e.g., energy drink, milk, cola) has been used as the focal good assigned to all students (e.g., background for cola provided in Elmore 2013). To illustrate an evaluation of externalities over the life cycle of a good, a sample assessment table is provided in ESM-C for bottled water (for background, see a video for the life cycle of bottled water in ESM-B). This example can be used as a case study to share with students as part of the introduction and research preparation while also providing an opportunity to introduce the Assessment Table (ESM-C). In reviewing the example, the instructor will have the opportunity to discuss and provide an example of assessing a good's relative life cycle impact using a 0-to-3 qualitative scale to the Assessment Table categories (where 0 is no impact, 1 is minimal, 2 is some impact, and 3 is significant impact; see ESM-C). (In the author's classes, bottled water became the guide for students' assessments of their chosen beverages. The evaluation and discussion of the externalities and market price of bottled water assisted students in qualitatively assessing the externalities specific to their chosen beverage to determine whether their research supported the market price of their chosen beverage.)

To assist students with their individual evaluation of the good chosen for the assignment, the instructor can provide students with a blank assessment table (ESM-C) and an assignment guide (ESM-D) that includes research resources and questions that align with the learning outcomes of the assignment. The questions should assist students in their determination of whether the observable market price is indicative of the externalities being assessed. Additionally, given the group discussion component of the assignment, the instructor should be prepared to provide class time to establish student self-selected working groups that can meet and facilitate individual research prior to the next class meeting dedicated to the assignment. At the second class meeting, the instructor should have copies of the assessment table (ESM-C) available for each group to populate with consensus (group) attributions of the individually assessed categories. During each group's debrief of their good assessment and price evaluation with the class, the instructor should be prepared to foster discussion related to the assessment by engaging the class in a discussion of the question prompts provided in the assignment guide (ESM-D). The conclusion of the assignment is expected to promote an understanding of drivers of demand and the relationship between excluded costs (externalities) and the observable market price of a good. At the conclusion of the second class, the instructor should be prepared to help the class think about how consumption could contribute to sustainable outcomes if the drivers of demand included an assessment of production externalities.

Activities

The assignment has four parts that are described below: (1) introduction facilitated by the instructor; (2) in-class research preparation by establishing student groups and completing initial work; (3) outside-of-class research; and (4) group discussions, class presentations, and concluding reflections.

Part 1: Introduction and Background Discussion (For Details, See ESM-A) (~20 min)

1. Introduce the general economics concepts addressed in Box 22.1.
2. Introduce students to the concept of negative externalities, specifically addressing water footprint, carbon footprint, human health impact, and natural resource impacts.
3. Provide example of research resources for determining a product's ecological footprint and health impacts (see Externalities in ESM-A) and show students how to find additional Internet resources with searches about the following topics in relation to the chosen good: water footprint, carbon footprint, health impact, and natural resource depletion.

Part 2: In-Class Preparation for Research (~30 min)

4. Introduce the assessment table (ESM-C) (~5 min). Establish (~5 min) a baseline footprint and human health impact using bottled water (ESM-C).

    (a) Distribute blank copies of the assessment table to students.
    (b) Explain the qualitative attribution of 0–3 for assessing the relative impacts of a good. Note that the attribution is determined after the research is collected for the assessment table categories. (The 0-to-3 attribution will likely be different for each student and this is fine; students will be able to discuss their findings in the group component of the exercise to arrive at a consensus.)
    (c) Convey that students are to use their overall assessment of the product to evaluate whether the market price adequately captures the negative externalities of the product (see the example in ESM-C). Prompt students to answer these questions: Is the price indicative of the externalities assessed, or should the market price be higher or lower? In other words, is the product fairly priced, or is there evidence of environmental and social subsidization? How does this factor in a fair price assessment of the product?

5. Provide time (~5 min) for students to self-select into groups of three or four based on a chosen good (from an assigned category such as a beverage or students' individual choices) that they all consume or are interested in. Explain that the purpose of the group is to collaborate on research about the life cycle of the

good within the context of four categories evaluated: water footprint, carbon footprint, human health impact, and natural resource impact. Inform students that the assignment relies on the use of factual data, but given the qualitative scaling of impacts, there will be a degree of subjective attribution to the price that should be able to be articulated and justified by the student.

6. Ask students to discuss (~13–15 min) why they have been consuming their chosen good (i.e., their consumption motivation) and have each group provide a brief summary (~3–5 min) to the class. Motivation may reflect price, preference, or taste. As part of the discussion, as needed depending on students' responses, the instructor can also highlight how marketing and advertising may have played a role in their consumption choices or how they may have inherited a consumption choice due to historical marketing and advertising success, i.e., their choices are behavioral legacies.

7. Provide students with the research assignment (ESM-D) to facilitate outside-of-class research and completion of the assessment table.

Part 3: Outside-of-Class Research

8. Give students ample time (e.g., 2 weeks) to individually research and find online resources and information on externalities, production processes, and other data for their good, leaving groups to determine whether they will meet outside of class or wait to share findings in the second class meeting set aside for the assignment.

Part 4: Group Discussions, Class Presentations, and Concluding Reflections (~50 min)

9. In the second class meeting, students should meet with their group, review findings and resources, and populate a group assessment table (ESM-C) with consensus assessments and research-based justifications (~25 min). The group assessment table will be the resource used in the group debrief of the assessment of their good with the class. The discussion questions from the assignment (ESM-D) can be employed to facilitate group discussion and serve as prompts to ensure that proper justifications have been provided for the assessment of externalities and ultimately the comparison of market price in relation to assessed externalities.

10. Groups should then share their market price assessment and rationale with the class (~3–5 min per group). It is recommended that each group has a spokesperson that provides a summary of the group's overall assessment of the validity of the good's market price relative to its life cycle externalities and resources used in determining and justifying their evaluation. Even if there is some redundancy among groups in the good chosen, in the author's experience, each group is able to share unique points due to the variation in research approaches.

11. Following the presentations, the instructor should engage the class in an evaluation of the assignment specific to the student's behavior to help them understand the responsibility inherent in their consumption behavior. How has the assignment affected your understanding of your own consumption behavior?

Will you modify your consumption behavior as a result of what you have learned? In concluding the assignment and discussion, the instructor can then direct the class to think about how prices and consumption could contribute to sustainable outcomes if prices and consumption decisions included externalities as part of more complete life cycle analyses of goods.

Follow-Up Engagement

- Students can be assigned readings (e.g., see Daly 1996; Heinberg 2011; Martenson 2011) and asked to evaluate how their individual consumption behavior can influence sustainability relative to planetary resources. For suggestions related to implementing the assignment, see Venkatesan (2015).
- Students can be directed to write a reflection essay, using references, to answer the questions provided in ESM-D. Since the reflection paper will, in part, be an outcome of the student's research for the assessment table, this should be provided with submission of the essay. An example rubric for evaluating this essay is provided in ESM-E.

Connections

- Awareness of the economics of sustainability can assist students in connecting and assessing causal economic relationships that have led to unsustainable environmental outcomes such as climate change due to carbon dioxide and methane production and habitat loss (e.g., deforestation) leading to species endangerment and extinctions (e.g., Chap. 18, this volume).
- Understanding the significance of consumption behavior with respect to sustainable outcomes can promote multidisciplinary interest in anthropology, sociology, and psychology to better understand the basis of societal and human behaviors (e.g., Chaps. 6, 21, 33).
- The study of history can provide a contextual foundation of sustainability from the perspective of the formation of generational behaviors and economic outcomes and provide the context for current "inherited" consumption behaviors that may result from prior time periods' marketed demands.

References

Daly HE (1996) Beyond growth: the economics of sustainable development. Beacon, Boston, MA
Daly HE (2014) From uneconomic growth to a steady-state economy. Edward Elgar, Northampton, MA

Dietz R, O'Neill DW (2013) Enough is enough: building a sustainable economy in a world of finite resources. Berrett-Koehler Publishers Inc., San Francisco, CA

Elmore BJ (2013) Citizen Coke: an environmental and political history of the Coca-Cola Company. Enterprise Soc 14:717–731

Fischer J, Dyball R, Fazey I et al (2012) Human behavior and sustainability. Front Ecol Environ 10:153–160

Greer JM (2011) The wealth of nature: economics as If survival mattered. New Society, Gabriola Island, BC

Hards S (2011) Social practice and the evolution of personal environmental values. Environ Val 20:23–42

Heinberg R (2011) The end of growth: adapting to our new economic reality. New Society, Gabriola Island, BC

Martenson C (2011) The crash course: the unsustainable future of our economy, energy, and environment. Wiley, Hoboken, NJ

Sen A (2010) Adam Smith and the contemporary world. Erasmus J Phil Econo 3:50–67

Venkatesan M (2015) Sustainability in the curriculum and teaching of economics: transforming introductory macroeconomics. Am J Educ Res 3:5–9

Venkatesan M (2016) Economic principles: a primer, a foundation in sustainable practices. Kona Publishing, Mathews, NC

Chapter 23
Business Sustainability and the Triple Bottom Line: Considering the Interrelationships of People, Profit, and Planet

Marilyn Smith

Introduction

Many types of organizations, including government, nongovernmental, nonprofit, and for-profit businesses, must make sustainability decisions about the sometimes-conflicting priorities of care for the planet, various interests of stakeholder groups, and economic or money issues (Farley and Smith 2014). The teaching activity described in this chapter introduces students to these complex interrelationships from the perspective of the for-profit business or commercial world. Traditionally, businesses referred to the bottom line as profit, but today they must consider the triple bottom line of profit, people, and planet, while navigating the differing priorities of their stakeholder groups ("Triple Bottom Line" 2009; also see Chaps. 2 and 24, this volume). In this context, Heizer and Render (2014) define business sustainability as "thinking not just about (the future of) environmental resources but also employees, customers, community, and the company's reputation." Other business texts use the same or similar definitions (e.g., Carroll and Buchholtz 2015, Russell and Taylor 2011).

Electronic supplementary materials: The online version of this chapter (doi:10.1007/978-3-319-28543-6_23) contains supplementary material, which is available to authorized users.

M. Smith (✉)
College of Business Administration, Winthrop University, Rock Hill, SC, USA
e-mail: smithm@winthrop.edu

© Springer International Publishing Switzerland 2016
L.B. Byrne (ed.), *Learner-Centered Teaching Activities for Environmental and Sustainability Studies*, DOI 10.1007/978-3-319-28543-6_23

For a company to remain successful in the long run (to be financially sustainable), it must be able to make a profit, by selling its products or services for more than they cost to produce and deliver. However, in context of the triple bottom line, this depends on sustainable relationships with people, i.e., stakeholders including employees, customers, shareholders (owner or investors), regulatory agencies, activists, and special interest groups (e.g., environmentalists), as well as the community where they operate. In a larger context for environmental sustainability, the company should not do harm to the planet and understand that most resources are limited. However, people, profit, and planet should not be considered separately, because they are interrelated, since a decision regarding one of them impacts the other two. Hence, balancing these three aspects for a company is challenging, especially the seemingly competing priorities of the different stakeholder groups.

The goal of the activity described below is to introduce students to the triple bottom line through an analysis of how a product's packaging change impacts the cost of the product, environmental issues related to the packaging, and therefore the business owner, customers, and communities. The quick and simple activity does not require prior business knowledge and could be used with students who are more familiar with the scientific and environmental perspectives of sustainability. Similarly, it could expand the perspective of business students with no environmental science background, who may think that the only objective of a corporation is to make a profit, by introducing them to other dimensions of a sustainable bottom line.

Learning Outcomes

After completing this activity, students should be able to:

- Define the three components of the triple bottom line and describe examples of each.
- Explain the relationships and trade-offs among people, profit, and planet in the design, manufacture, and use of a product, packaging, or service.

Course Context

- Used as an introductory lesson in a junior-level required college business course with 12–45 students
- 10–15 min in one class meeting, or can be expanded with additional provided case studies
- Can be used in any general course that discusses sustainability, so that business students learn about environmental aspects and other students learn the business aspects. No specific background is expected

Instructor Preparation and Materials

Savitz (2006) gives a good introduction to the triple bottom line. If the instructor does not have business experience, additional general background information and key journal articles are provided in ESM-A and ESM-B, respectively. Also, four cases and further talking points about the imperative to balance issues related to people, profit, and planet rather than to prioritize one to the detriment of the others are suggested in ESM-C. The first two cases could be used as introductory material or could be used instead of the jar example provided below. The Nike and Patagonia cases are more complex and could be used as follow-up activities.

Before class, the instructor needs to identify and obtain an example product or service that illustrates the triple bottom line through a design change targeted toward sustainability. (A photo alone could also work for the activity.) For example, a peanut jar (Fig. 23.1) says the packaging has been reduced by 84 % by changing from glass to plastic to a thinner, yet stronger, plastic. Another example would be a T-shirt made from recycled plastic bottles. The chosen product should enable discussion about the following two key points that the instructor should be prepared to present to or discuss with students:

- To be successful over the long term (sustainable), companies must consider not just profit, but also people, and the planet.
- People, profit, and planet are connected. Companies need products and services that can be produced and sold for a profit, while respecting people and honoring the limits of the planet.

Activities

First, students should be provided a brief (2–5 min) introduction of the business definition of the triple bottom line (see introduction, references cited above and in Electronic Supplementary Materials for more background).

Then, after the selected focal product or service has been introduced, the students can start to brainstorm: how does this product or service change impact people, profit, and planet? The instructor may prompt the students with additional questions or provide an additional example or two to help them start (see Table 23.1 for questions and considerations for the peanut jar example (Fig. 23.1)). The brainstorming could be open to the whole class at once, or students could be asked to write their own notes for a specified time (e.g., 3 min) and then discuss with a partner or small group, before sharing responses with the whole class. One part of the triple bottom line can never be prioritized at the complete expense of the other two over the long run. Companies must make a *reasonable* profit to meet the requirements of owners/investors/stockholders, but the goal of maximizing profit at the expense of the environment or other stakeholders will not be sustainable over time.

Fig. 23.1 A packaging
change for a peanut jar
provides an example to
discuss impacts on people,
profit, and the planet. (This
image, in color, is provided
in ESM-A for instructor
use).

Follow-Up Engagement

- Students could play the online Energy Challenge Game, created by Duke Energy, the largest electric provider holding company in the USA, to educate consumers about the challenges they face in balancing people, profit, and planet while deciding which energy sources to use (available at: http://energychallenge.duke-energy.com). Players decide how Duke should increase its capacity to meet forecasted demand, by either adding plants (coal, nuclear, wind or solar), upgrading current plants to either increase capacity or reduce emissions, or closing plants. As decisions are made, the game shows the results in terms of the effect on capacity, CO_2 emissions, and costs and tracks these three results as "scores."
- ESM-C includes four additional case studies in the order of simplest to most complex that can be used to illustrate the interrelatedness of people, profit, and planet.
- Related activities about the triple bottom line are provided in Chaps. 2 and 24, this volume.

Connections

- Some companies now include business sustainability under the umbrella of corporate social responsibility, a much broader concept, including ethics and philanthropy. Carroll and Buchholtz (2015, p. 28) define corporate social responsibility as: "…seriously considering the impact of a company's actions on society" which "requires the (business) individual to consider his or her acts in terms of a whole social system, and holds him or her responsible for the effects of his or her acts anywhere in that system." A related question to discuss with students is: What is the relationship between business sustainability and ethics?
- Allen (2009) defines greenwashing as "the act of misleading the public regarding the environmental practices of a company or the environmental benefits of a product, service, or business line" and lists (as of 2009) the ten worst US offenders of greenwashing. Laufer (2003) provided nine examples of how companies greenwash.

Table 23.1 Example questions and triple bottom-line considerations

Question/issue	Considerations
How does less packaging impact costs to the company?	• Reduced costs • More profit • Satisfied owners/investors
How would the packaging change possibly impact customers?	• Savings possibly passed on to the consumer • Easier to carry and handle
How would the lighter weight impact transportation?	• Savings in fossil fuel used • Less cost to transport • Less transportation emissions

What are the recycling issues related to the package change?	• Is plastic more likely to be recycled than glass or vice versa?
	• What are the people, planet, and profit issues related to recycling glass or plastic?
	  – How are the recycling processes different?
	  – What are costs of recycling the respective materials?
	  – What happens if glass goes to the landfill?
	  – What happens if plastic goes to the landfill?

What are impacts of this change on workers?	• Easier for workers at the factory, warehouses and distribution centers, and stores to lift and handle
Key summary/big picture question: What if the company only made the change to save money? Would the answers above still be the same?	Note that the definition says "To be successful over the *long term*…." Companies can make short-term decisions based on profit, as long as there is no harm to others. But to be successful over the long term, we cannot destroy natural resources and show no concern for other stakeholders besides owners

To connect this concept with the activity, students can be asked whether a company can be successful by just pretending to be concerned about the planet and what the requirements should be for a product to be labeled as "green" or "eco-friendly."

- Although this lesson is approached from a business perspective, the triple bottom line can be discussed in context of ecological issues such as resource use and biodiversity conservation. For example, the whaling industry in the late 1800s provides a historical catastrophic example of pursuing profit without considering other concerns.
- The challenge of balancing competing interests is explored in Chap. 19, while more about the economics of sustainability is discussed in Chap. 22.

References

Allen A (2009) The "Green" hypocrisy: America's corporate environment champions pollute the world. http://247wallst.com/energy-business/2009/04/02/the-%E2%80%9Cgreen%E2%80%9D-hypocrisy-america%E2%80%99s-corporate-environment-champions-pollute-the-world/. Accessed 13 May 2015

Carroll A, Buchholtz A (2015) Business and society: ethics, sustainability, and stakeholder management, 9th edn. Cengage Learning, Stamford, CT

Farley H, Smith Z (2014) Sustainability: if it's everything, is it nothing? Routledge, New York, NY

Heizer J, Render B (2014) Principles of operations management: sustainability and supply chain management, 9th edn. Pearson, Upper Saddle River, NJ

Laufer W (2003) Social accountability and corporate greenwashing. J Bus Ethics 43:253–261

Russell R, Taylor B (2011) Operations management: creating value along the supply chain, 7th edn. Wiley, Hoboken, NJ

Savitz A (2006) The triple bottom line: how today's best-run companies are achieving economic, social, and environmental success- and you can too. Jossey-Bass, San Francisco, CA

Triple Bottom Line (2009) The economist http://www.economist.com/node/14301663/. Accessed 14 Jan 2015

Chapter 24
A Triple-Bottom-Line Analysis of Energy-Efficient Lighting

Timothy Lindstrom and Catherine Middlecamp

Introduction

The activity in this chapter focuses on an aspect of social comfort and convenience that is often taken for granted: artificial lighting. Lighting trends worldwide are shifting toward greater energy efficiency (Martinot and Borg 1998; Cook 2000). Incandescent light bulbs are being discarded in favor of more efficient compact fluorescent lights (CFLs) and light-emitting diodes (LEDs). The US federal government is facilitating this shift, having passed the Energy Independence and Security Act (EISA) of 2007 that raised lighting efficiency standards and phased out the manufacture of most incandescent light bulbs (One Hundred Tenth Congress 2007). As energy-efficient lighting becomes the new norm, payoffs are likely to be seen through reductions in utility bills, energy consumption, pollution, and carbon emissions (Center for Climate and Energy Solutions 2011).

Institutions of higher education are embracing efficient lighting as part of their ongoing efforts to reduce the environmental impacts of campus operations (Rappaport 2008). At the same time, instructors are seeking opportunities to utilize their campuses as living-learning laboratories to explore sustainability issues (Schrand et al. 2013). By integrating a campus lighting project into the classroom, instructors can partner with staff from campus operations to develop beneficial partnerships for campus sustainability and project-based learning (Herman Miller, Inc. 2005).

Electronic supplementary materials: The online version of this chapter (doi:10.1007/978-3-319-28543-6_24) contains supplementary material, which is available to authorized users.

T. Lindstrom (✉) • C. Middlecamp
Nelson Institute for Environmental Studies, University of Wisconsin-Madison,
Madison, WI, USA
e-mail: timothy.lindstrom@wisc.edu; chmiddle@wisc.edu

To this end, the activity described here evaluates the sustainability of a campus lighting project using the framework of the triple bottom line which emphasizes benefits for the economy, environment, and community (Slaper and Hall 2011; also see Chaps. 2 and 23, this volume). Students calculate the payback period of a small lighting project (using the provided case study or one from their campus), estimate its effects on associated carbon dioxide (CO_2) emissions, and reflect on its societal impacts. The overarching goals are for students to (1) describe how sustainability relates to everyday decisions on their campus and (2) connect lighting and energy efficiency to larger economic, environmental, and societal considerations.

Learning Outcomes

After completing this activity, students should be able to:

- Compare and contrast the energy efficiency of incandescent, CFL, and LED light bulbs.
- Calculate the costs over time and payback periods for using different models of light bulbs.
- Estimate the CO_2 emissions attributed to electricity use for lighting.
- Describe the relationships among energy use, economic costs, and CO_2 emissions.
- Apply the triple-bottom-line perspective to lighting choices and to decision-making on personal and institutional levels.

Course Context

- Developed as part of an energy unit for an introductory environmental studies course with 30–40 undergraduate students
- 45–60 min in one class period
- Before the activity, students should have an understanding of the concepts of power, energy, and the triple bottom line. Electronic Supplementary Materials, written specifically for this purpose, are provided that contain introductory text and practice problems
- Adaptable to courses of any class size or level that include discussions of energy and energy efficiency, sustainability, and environmental economics

Instructor Preparation and Materials

Instructors can implement this activity in one of three ways:

1. Students evaluate the University of Wisconsin-Madison lighting project as a case study (the Union South Marquee) using data provided in student and instructor files (ESM-A and B). In this case, the project replaced 75-W halogen lamps with 13-W LEDs in 11 ceiling fixtures at a cinema in one of the student unions.
2. Students evaluate a lighting project on their own campus. To do this, the instructor must obtain data for a completed lighting project, with assistance as needed from campus operations and facility personnel or a sustainability director. The necessary data are:

   - The model of each bulb (e.g., incandescent, halogen, CFL, LED)
   - Individual cost of each model
   - Factory lifetime in hours of each model
   - Wattage of each model
   - Electricity rate in \$/kWh charged to the institution by the electric utility
   - Labor costs for the project
         The environmental assessment requires a conversion factor for mass of CO_2 emissions per unit of electricity. Contact a local utility to obtain this value. Alternatively, the most recent EPA estimates by region are available from the EPA eGRID webpage (EPA 2014a, b). Search for *2010 eGRID subregion GHG output emission rates* (EPA 2014a). Use the appropriate region's CO_2 value under "Annual total output emission rates" and convert CO_2 from lb/MWh to lb/kwh for a more manageable value (divide lb/MWh by 1000).

3. If actual campus lighting data are unavailable, students can evaluate a hypothetical lighting project designed by their instructor. For example, the instructor could obtain an incandescent light bulb and an equivalent CFL or LED and create a lighting upgrade scenario using data from the "Lighting Facts" on the packaging that provide everything except the local electricity rate and labor costs. The electricity rate can be taken from the instructor's residential utility bill. Labor costs in this case could be taken from the case study in ESM-A (p. 5).

Instructors should form student groups of two or three and ensure that each group will have a calculating device. For demonstration purposes, instructors should acquire either actual light bulbs or photographs. Instructors should review ESM-A and B and as needed edit the text and data to fit their own lighting project. Both documents provide detailed instructions for how students should work through the triple-bottom-line assessment of the lighting project. Prepare and print a copy of ESM-A for each student.

Activities

1. Introduction and warm-up (10–15 min)

   (a) A discussion helps prepare students for this activity. A list of prompts is provided in ESM-B. Instructors may address these prompts in an all-class discussion or allow students to respond in small groups first.

(b) ESM-A includes two pre-activity survey questions (pp. 3–4) that may be used in tandem or in place of the questions from the student worksheet.

(c) Transition into the activity by explaining to students that they will be analyzing the project from three different perspectives that follow the triple bottom line. This is a good opportunity to reiterate what the triple bottom line is and how it is applied to decision-making.

The following sections closely follow ESM-A. All page references refer to this file.

2. Economic analysis (15–20 min)

This analysis examines the trade-offs between short-term and long-term costs of efficient lighting (p. 5). Students should discover that the initial investment in more efficient lighting soon pays for itself and can lead to significant savings over the lifetime of the bulbs.

(a) Divide the class into groups of two or three. Each student should have a printed copy of the handout (ESM-A), and each group should have a calculating device.

(b) Using the data in Table 1 (p. 5), each group will perform a series of calculations, eventually arriving at a simple payback period for the lighting upgrade. Estimates are given for three lengths of time that the lights are on each day: 6, 10, or 14 h. Instructors should assign one of the three values to each group. Groups will compare their results later in the activity.

(c) The handout walks students step-by-step through an initial set of calculations (p. 5). These calculations are not the payback period; rather, they are the elements needed to then calculate the payback period. Details of the specific calculations are provided in ESM-B (p. 16).

(d) Students are challenged to calculate the payback period of the lighting project using the lighting data and the results of their initial calculations. Instructors are encouraged to remain fairly hands-off in favor of allowing the students to work through this process.

(e) Groups compare their results with others who were assigned a different daily use estimate, reflecting on how the payback period is affected by how long the lights are on. This step is explicitly written into ESM-A to facilitate compliance (p. 8).

(f) Students answer follow-up questions and draw conclusions as to whether or not more efficient lighting is economically sustainable (p. 8).

3. Environmental analysis (10–15 min)

This analysis helps students make connections between the light switch that is flipped on and the power plant that supplies the electricity (p. 9). By converting electricity use into mass of CO_2, students quantify and compare an environmental impact across two scenarios of efficient and inefficient lighting.

(a) Groups remain unchanged and use the same daily use estimate (6, 10, or 14 h) assigned to them in the economic analysis.

(b) This analysis requires a conversion factor for mass of CO_2 per kwh of electricity. For information about obtaining this value, see the "Instructor Preparation and Materials" section.

(c) Students calculate the annual reduction in CO_2 emissions from the lighting project.

(d) Groups compare their result with others who were assigned a different daily use estimate, exploring how CO_2 emissions are affected by lighting preferences and behaviors.

(e) Students respond to follow-up questions and evaluate whether more efficient lighting is environmentally sustainable (p. 10).

4. Societal analysis and reflection on the triple bottom line (10–15 min)

This analysis reflects on the societal impacts of the lighting project, asking students to think about how efficient lighting can improve the well-being of communities (p. 11).

(a) The handout challenges students to come up with at least one example of societal impacts for each of two different scales (ESM-B provides examples):

- The narrower-in-scope and immediate impacts relate to how the lighting project will affect the building and its inhabitants (faculty, students, and staff).
- The broader, long-term impacts relate to how efficient lighting will affect public health and societal well-being, including impacts at local/regional (better air quality and associated health and economic benefits) and global scales (climate change mitigation through reduced greenhouse gas emissions).

(b) Students conclude the activity by aggregating their findings into a brief summary of the sustainability of the lighting project (p. 11). This summary should include an analysis of each aspect of the triple bottom line and should argue whether or not the project is a "win" for overall sustainability. If time is short, this final task may be a homework assignment. See *Follow-Up Engagement* for more details.

5. Conclusion (10–15 min)

(a) Ask groups to discuss with the class their main findings for the economic, environmental, and societal analyses above.

- What are their most significant takeaways from the activity?
- Did the activity affect their views about efficient lighting in particular or energy use in general?

(b) Instructors should prompt students to reflect on how the estimates of daily lighting use influenced both the payback period of the project and the annual reduction in CO_2 emissions. Display student results on the board for comparison. Do they consider this influence to be significant?

(c) Ask students to volunteer their opinion on the sustainability of the lighting project. Push students to justify their opinion from each aspect of the triple bottom line, and tease out differing opinions if they exist.

(d) Instructors who wish to extend the discussion may refer to the list of questions provided in ESM-B.

Follow-Up Engagement

- In a brief (1–2 page) reflective essay, ask students to evaluate the sustainability of efficient lighting, addressing each aspect of the triple bottom line and referencing results from the activity. If appropriate, require students to use references to support their conclusions.
- Ask students to identify a light bulb in their home or residence hall that could be replaced. Have them search online for the manufacturer, wattage, and model of the existing bulb (e.g., Philips 60-W incandescent bulb). Then ask them to find an equivalent high-efficiency bulb either online or at a store. Following a procedure similar to the one in this activity, have students estimate the payback period for purchasing the new light bulb.
- Assign students to read (e.g., Manley 2013; Truini 2012), or conduct their own research, about issues and arguments surrounding EISA and organize an in-class debate. The government's decision to phase out incandescent light bulbs is not universally popular; on the contrary, it's a contentious topic that is fairly split along party lines (Herrman 2011; Stika 2015). The political left tends to support EISA as the right step for curbing energy consumption, whereas the political right tends to criticize EISA as another example of "big government" intruding on people's lives by eliminating personal choices (Walsh 2011).

Connections

This activity connects to other environmental and sustainability study topics, including:

- Green engineering and design. How do we make new construction and remodeling projects more environmentally friendly? What are strategies for reducing energy use, water use, and waste during the construction and use of a building?
- Environmental psychology and sociology. How do people make decisions when there are temporal trade-offs between economics and the environment (e.g., the LED is more expensive to buy, but it's better for the environment and may save money in the long run)? (See Chaps. 6, 19, 20–23, this volume, for related topics and activities.)
- Environmental policy and politics. The EISA legislation is an example of the minefield that elected officials cross when making decisions about environmental regulations. Where does conflict arise between environmental policymaking and individual liberties? What role does and should government play in legislating environmentalism?

- Environmental impacts of energy use. How does the energy system affect air quality, criteria pollutants, greenhouse gas emissions, and climate change? Where do renewables, energy efficiency, and consumption patterns fit in as sustainable solutions? The choices we make now and in the future will have significant impacts on the health of our societies, ecosystems, and planet. (For related activities, see Chaps. 25, 26, 32 and 39, this volume.)

Acknowledgments Funding to develop this activity at the University of Wisconsin-Madison was provided by WE CONSERVE and the UW-Madison Office of Sustainability.

References

Pew Center on Global Climate Change (2011) Center for climate and energy solutions. Lighting Efficiency. http://www.c2es.org/docUploads/LightingEfficiency.pdf. Accessed 18 Jan 2015

Cook B (2000) New developments and future trends in high-efficiency lighting. Eng Sci Educ J 9(5):207–217

Herman Miller, Inc (2005) Creating a culture of sustainability: how campuses are taking the lead. http://www.hermanmiller.com/hm/content/research_summaries/wp_Campus_Sustain.pdf. Accessed 7 Dec 2014

Herrman J (2011) Is the light bulb ban a bright idea? In: Popular mechanics. http://www.popular-mechanics.com/technology/gadgets/a7210/is-the-light-bulb-ban-a-bright-idea-6459591/. Accessed 5 Mar 2015

Manley N (2013) Incandescent light bulbs: the controversy. In: Del mar fans & lighting. http://www.delmarfans.com/educate/basics/illegal-incandescent-light-bulbs/. Accessed 16 Feb 2015

Martinot E, Borg N (1998) Energy-efficient lighting programs: experience and lessons from eight countries. Energ Pol 26(14):1071–1081

One Hundred Tenth Congress of the United States of America (2007) Energy independence and security act of 2007. U S Government Printing Office. http://www.gpo.gov/fdsys/pkg/BILLS-110hr6enr/pdf/BILLS-110hr6enr.pdf. Accessed 27 Sep 2013

Rappaport A (2008) Campus greening: behind the headlines. Environ Sci Pol Sust Dev 50:6–17

Schrand T, Benton-Short L, Biggs L et al (2013) Teaching sustainability 101: how do we structure an introductory course? Sust J Record 6:207–210

Slaper TF, Hall TJ (2011) The triple bottom line: what is it and how does it work? In: Indiana business review. http://www.ibrc.indiana.edu/ibr/2011/spring/pdfs/article2.pdf. Accessed 21 Feb 2015

Stika N (2015) The great light bulb ban that wasn't. In: Council of smaller enterprises. http://www.cose.org/About%20COSE/News%20and%20Media/Blog/The%20Great%20Light%20Bulb%20Ban.aspx. Accessed 20 Dec 2014

Truini J (2012) What you need to know about the light bulb law. In: Popular mechanics. http://www.popularmechanics.com/home/how-to/a8453/what-you-need-to-know-about-the-lightbulb-law-14789203/. Accessed 5 Mar 2015

(EPA) U S Environmental Protection Agency (2014a) eGRID. Clean Energ Resour. http://www.epa.gov/cleanenergy/energy-resources/egrid/index.html. Accessed 18 Oct 2013

(EPA) U S Environmental Protection Agency (2014b) eGRID. Clean Energ Resour. http://www.epa.gov/cleanenergy/documents/egridzips/eGRID_9th_edition_V1-0_year_2010_GHG_Rates.pdf. Accessed 18 Oct 2013

Walsh SC (2011) House votes to hamper a law on light bulbs. New York Times. http://www.nytimes.com/2011/07/16/business/house-votes-to-withhold-funding-for-light-bulb-law.html. Accessed 15 Mar 2015

Chapter 25
Go with the Flow: Analyzing Energy Use and Efficiency in the USA

James D. Wagner

Introduction

The extraction and use of various forms of energy are at the nexus of many environmental issues. Although we live in an age of diverse energy sources—fossil fuels, nuclear, solar, wind, and geothermal—most developed countries still rely heavily on burning the remains of ancient organisms to fuel their economies. In 2011, fossil fuels represented 83.6 % of the energy consumed by the USA, 85.1 % for the UK, 79.8 % for Germany, 48.6 % for France, 88.3 % for China, and 94.8 % for Australia (World Bank 2015). It is difficult to appreciate the scale and scope of energy use at the national level, but this understanding is critical if we are to have informed and productive conversations about our future energy policies and the potential for renewable energy solutions (Jacobson and Delucchi 2011; Murray and King 2012; Yuan et al. 2014; World Energy Council 2015).

To initiate those discussions with students, the exercise described below uses a visually stunning chart that is produced every year by the US Lawrence Livermore National Laboratory (https://flowcharts.llnl.gov/). The energy flow chart summarizes how various energy sources are used by different sectors of the economy and enables students to calculate how efficiently those sectors use energy. Resources for energy use in other countries are provided in Electronic Supplementary Material (ESM) A. Through calculating proportions, this exercise reinforces to students that (1) the USA relies heavily on fossil fuels; (2) over 2/3 of the energy consumed in the

Electronic supplementary materials: The online version of this chapter (doi:10.1007/978-3-319-28543-6_25) contains supplementary material, which is available to authorized users.

J.D. Wagner (✉)
Biology Program, Transylvania University, Lexington, KY, USA
e-mail: jwagner@transy.edu

USA is for electrical generation, industrial, and transportation needs; (3) transportation and electrical generation are dramatically energy inefficient; and (4) not all energy sources have equal application and use. A goal of this exercise is to fuel spirited and informed student-led discussions about the roles of sustainable energy sources in our economy and the scale and scope of the problem our society faces in moving away from our reliance on fossil fuels (also see Chap. 26, this volume). In addition, this exercise helps give students a strong foundation for exploring issues associated with energy extraction and use, such as mountaintop removal for coal mining, pollution (e.g., Lamborg et al. 2014, Chap. 32, this volume), and effects of global climate change.

Learning Outcomes

After completing this activity, students should be able to:

- Interpret the energy flow chart produced by the Lawrence Livermore National Laboratory.
- Compare and contrast how various sectors of the US economy rely on specific energy sources.
- Calculate and rank energy efficiencies for various sectors of the US economy (electrical production, residential, commercial, industrial, and transportation).
- Calculate the relative reliance of the US economy on fossil fuels compared to other energy sources.

Course Context

- Successfully used in a nonmajor college-level environmental science course with 20–24 students
- Can be completed in 50 min but can be stretched to a longer class period by encouraging discussion or asking the students to project future changes in the types of energy used
- No student background knowledge required beyond basic math skills for calculating proportions

Instructor Preparation and Materials

Preparation for this exercise is minimal, but instructors who wish to prepare with additional background can consult the additional reading list in ESM-A which also includes discussion prompts for use with students. Instructors should understand the energy flow chart's logic and organization, the types of energy sources presented (which are self-evident by their names), and how each sector is defined. The diagram from 2013 (provided in ESM-C) is used as an example in the description below and for the calculations explained in ESM-A. The most recent and previous years' diagrams can be found at https://flowcharts.llnl.gov/archive.html#energy_archive or located using the keywords "Energy Flow Charts" at https://www.llnl.gov/. Although the flow charts may be presented in units of quads or exajoules (American Physical Society 2015), understanding the specifics of the units is not required for this exercise since all calculations are based on percentages for ease of comparison and evaluation.

Providing a color copy of the flow chart for each student is ideal but may be cost prohibitive. Printing the flow charts onto color transparencies or laminating printed copies that can be reused by different classes may be an affordable alternative. Since the flow charts made since 2003 utilize bold distinctive colors with the numbers used in the exercise clearly identified in black, projecting the flow charts can be an acceptable alternative in classrooms with computer projectors. A copy of the worksheet (ESM-B) and a simple calculator are needed by each student to complete the activity. The worksheet can be edited as needed to fit within the time constraints of the class by removing or adding questions or using some as homework problems.

Activities

One of the major themes of this exercise is that key changes must be made in transportation and electricity generation if there is any chance to make substantial reductions in greenhouse gas emissions in the USA. Energy efficiency improvements in residential and commercial sectors will help but are trivial compared to the energy consumption and pollution from transportation and electricity generation.

1. Introduce the exercise by discussing the energy flow chart's composition and logic. The flow chart is analogous to a utility bill for the entire USA in that it reveals the total annual energy used by all Americans and shows how they used that energy in their homes, transportation, and commercial enterprises (~5 min).
2. Group students in pairs or triplets supply them with a copy of the chart and ensure that each group has a calculator.
3. Orient the students to the top of the chart where it indicates total US energy use (e.g., for 2013 it is 97.4 Quads; see ESM-C). This will be the denominator in the calculations for Tables 1 and 2 on the student worksheet (ESM-B). Ask students to fill out Table 1. Have them compare their answers with other groups (8–10 min). If desired, this is a good time to engage students in discussion with the following questions (suggested answers for these and the ones below are provided in ESM-A):

    (a) How much does the US economy rely on fossil fuels to operate?

    (b) How do we define renewable and conventional energy sources and carbon and carbonless energy sources (EPA 2015a)?

    (c) Why is most of the renewable energy used in the generation of electricity (NREL 2015)?

4. Orient the students to the boxes which indicate total energy use *within* each sector: electrical generation, residential, commercial, industrial, and transportation.

    (a) To allow the totals to add to 100 % in Table 2, the amount reported in the Electrical Generation box must be adjusted by subtracting out the amount of electrical energy that is then distributed into the consuming sectors (Residential, Commercial, Industrial, and Transportation); see the answer sheet in ESM-B for a detailed explanation. After 5–8 min, have the students compare their answers with other groups. If desired, additional discussion questions could be integrated here:

- Which sectors of the US economy use the most and least energy?
- What sources of energy seem to go predominantly to each of the sectors?
- How does this method actually underrepresent the energy required for electrical generation?

5. Table 3 explores energy inefficiency in the US economy. Within the charts, wasted energy is labeled as *rejected energy* and is represented by the pale gray stream that converges to the top right side of the chart. (See ESM-A for extended discussion of rejected energy and how it is determined.) Completing the calculations for Table 3 requires students to notice that the amount of energy lost (rejected) within each sector is represented by the value above the pale gray stream to the right of the box associated with that sector. Using the total amount of energy used within a sector as the denominator and the value associated with rejected energy for that sector as the numerator, students can calculate energy inefficiency per sector (e.g., in 2013 (ESM-C), electrical generation wasted 67.5 % of the energy used in the process of creating electricity (25.8 quads/38.2 quads)). Suggested discussion questions to include at this point are:

    (a) Why is transportation so inefficient? How can we improve efficiency?

    (b) Why is electricity so inefficient in its production?

    (c) Residential sector wastes about 35 % of the energy it requires. What aspects of a home contribute to this inefficiency?

6. To close the discussion, it is worth introducing the ideas of tax breaks, tax subsidies, and other ways in which governments encourage or discourage the use of energy. Discussion questions could be:

    (a) If you controlled government funding to reduce carbon emissions, on which sector should the efforts be focused to improve efficiency? Why?

    (b) If we wanted to improve transportation efficiency, what are the hurdles to improving that sector of our economy?

Follow-Up Engagement

- An easy way to extend this assignment is to have students compare the most recent year's flow chart with one from a previous year found in the archive section of LLNL's website (above). In the author's experiences, students are interested in the degree of changes within their own lifespan. (For such comparisons, note that the charts prior to 2003 merged residential with commercial sectors into a single entity which complicates direct comparisons to recent charts; however, there is little difference in efficiency between residential and commercial sectors so those earlier charts are still worth using.)
- Another exercise is to examine the State-by-State Energy Flow report of 2008 (Laboratory 2015) which has energy flow charts for each of the individual US states. Students may find it interesting to compare their home state energy use patterns with that of the entire USA.
- Assign students to role-play as politicians who are trying to increase efficiency of the transportation sector using either "carrots or sticks" (i.e., incentives and penalties) approach. Carrots include giving tax credits, rebates, or subsidies to individuals and corporations when they invest in energy efficient technologies. Sticks involve increasing taxes (e.g., gasoline or carbon tax) or fines (cap and trade) to encourage individuals or corporations to reduce their use of undesired energy sources. Have them debate which approach will be most effective.

Connections

- This exercise gives students a sense of the scale of energy use at the national level which allows them to be more informed during explorations of sustainable/renewable energy sources and their integration into our society (e.g., Chap. 26, this volume).
- Our reliance on coal to fuel our electricity production results in large-scale destruction of natural habitats through strip mining and mountain top removal for coal extraction. This disrupts ecosystems and communities, issues which relate to biodiversity conservation and social sustainability and environmental justice (e.g., Chaps. 27–30, this volume). Data from this analysis suggests it may be easier to reduce our reliance on coal since it is used within a single sector compared to other fossil fuels (natural gas and petroleum) that span multiple sectors.
- Coal accounts for only 39 % of the electricity generated in the USA, but it is responsible for 75 % of the CO_2 emissions in generating electricity. Electrical generation is the largest source of greenhouse gas emissions in the USA (EPA 2015b). (For an electricity-use activity, see Chap. 24, this volume.)

Acknowledgments The author would like to thank all of his students enrolled in his Environmental Science course at Transylvania University over the years who completed this exercise and helped to clarify and develop the assignment. I would also like to thank my colleague Dr. Josh Adkins for reviewing and giving feedback on an early draft of the chapter and Dr. Kathleen Jagger for making me aware of this book project.

References

American Physical Society (2015) Energy units. http://www.aps.org/policy/reports/popa-reports/energy/units.cfm. Accessed 24 Feb 2015

EPA (2015a) Green power market. http://www.epa.gov/greenpower/gpmarket/. Accessed 24 Feb 2015

EPA (2015b) Electricity sector emissions. http://www.epa.gov/climatechange/ghgemissions/sources/electricity.html. Accessed 10 Mar 2015

Jacobson MZ, Delucchi MA (2011) Providing all global energy with wind, water, and solar power, part I: technologies, energy resources, quantities and areas of infrastructure, and materials. Energ Pol 39:1154–1169. doi:10.1016/j.enpol.2010.11.040

Lawrence Livermore National Laboratory (2015) Energy flow. https://flowcharts.llnl.gov/. Accessed 24 Feb 2015

Lamborg CH, Hammerschmidt CR, Bowman KL et al (2014) A global ocean inventory of anthropogenic mercury based on water column measurements. Nature 512:65–68. doi:10.1038/nature13563

Murray J, King D (2012) Climate policy: oil's tipping point has passed. Nature 481:433–435. doi:10.1038/481433a

NREL (2015) Learning about renewable energy. http://www.nrel.gov/learning/. Accessed 24 Feb 2015

World Bank (2015) Fossil fuel energy consumption (% of total). http://data.worldbank.org/indicator/EG.USE.COMM.FO.ZS. Accessed 24 Feb 2015

World Energy Council (2015) World energy issues 2015. http://www.worldenergy.org/data/issues/. Accessed 24 Feb 2015

Yuan J, Xu Y, Zhang X et al (2014) China's 2020 clean energy target: consistency, pathways and policy implications. Energ Pol 65:692–700. doi:10.1016/j.enpol.2013.09.061

Chapter 26
Exploring Complexities of Energy Options Through a Jigsaw Activity

Nathan Hensley

Introduction

Currently, the world's transportation and commerce infrastructure is designed for coal, oil, and natural gas—energy resources, which are relatively easy to transport and store, can be used when desired and have low production costs and high energy output (Murphy 2012; Chap. 25, this volume). However, these fossil fuels are non-renewable and have significant negative side effects. Externalities, which are side effects of using energy sources that are not reflected in their price, are also a factor in considering the potential limitations of these energy sources. Examples include their impacts on human health and the natural environment from pollutants released by combusting the fuels (e.g., Chap. 32, this volume) and the noise pollution during the production processes.

Therefore, we are in an energy crisis. It is often argued that alternative energy sources (anything besides fossil fuels) should be used to help "solve" this crisis. However, factors such as economic and environmental costs and efficiency should be considered when looking at the possibility of transitioning to these alternatives. To be useful, an energy resource must provide more energy to society than is required to generate it. The term "net energy ratio" refers to the amount of energy generated by an energy source versus the amount of energy expended to make it usable (Heinberg 2009). However, even an energy source with a large and positive net energy ratio (one that has low resource investment but produces a lot of energy) can be essentially worthless if it cannot be accessed by the general population.

Electronic supplementary materials: The online version of this chapter (doi:10.1007/978-3-319-28543-6_26) contains supplementary material, which is available to authorized users.

N. Hensley (✉)
Department of the Environment and Sustainability, Bowling Green State University, Bowling Green, OH, USA
e-mail: nhensle@bgsu.edu

© Springer International Publishing Switzerland 2016
L.B. Byrne (ed.), *Learner-Centered Teaching Activities for Environmental and Sustainability Studies*, DOI 10.1007/978-3-319-28543-6_26

Helping students understand these energy issues is important because an energy-literate person is more likely to make well-informed decisions on topics ranging from energy conservation to utilizing energy-efficiency equipment on a day-to-day basis. As such, the goal of the activity described below is for students to learn about energy sources and their net energy ratios so that they can explain the advantages and disadvantages (including externalities) of each. This is achieved through the use of a jigsaw activity in which students learn assigned components of the content matter and are responsible for educating their classmates about the assigned components of the material.

Learning Outcomes

After completing this activity, students should be able to:

- Define the concept of net energy ratio and explain how it impacts the viability of an energy source.
- Articulate the advantages and disadvantages of using different energy sources.
- Describe what an externality is and list ones associated with using fossil fuels.
- Explain the challenges of replacing fossil fuels with alternative energy sources.

Course Context

- Developed for introductory-level environmental science, environmental studies, or sustainability studies classes, but appropriate for advanced classes
- Approximately 50 min in one class meeting
- No specific background knowledge is needed by students
- Adaptable for courses ranging from 16 to 36 students

Instructor Preparation and Materials

Before the day of the activity:

- The instructor should read Murphy (2012) about the energy sources, choose the ones that will be used in the activity (see Box 26.1), and review the completed "Energy Source Table" in Electronic Supplementary Materials (ESMs) A.
- Students should be assigned a reading pertaining to net energy ratio (Heinberg 2009, pp. 23–28). Heinberg uses the term "energy return on energy invested" (EROEI) interchangeably with net energy ratio, a point which may need to be emphasized when the assignment is given. Heinberg (2009) also contains sec-

Box 26.1: Directions for Creating the Jigsaw-Activity Groups

1. Place students into small groups of four to six. These are the "seed groups" (see Fig. 26.1).
2. Each student in a seed group becomes an "expert" on two of the following energy sources by reading about each in Murphy (2012): conventional oil (p. 6), unconventional oil (p. 8), natural gas (p. 9), coal (p. 11), nuclear (p. 12), hydropower (p. 13), geothermal (p. 14), liquid biofuels (p. 15), industrial wind (p. 17), and solar voltaic (p. 18). For students to answer question #8 in the discussion questions (ESM-D), liquid biofuels will need to be one of the energy sources assigned.
3. Form new groups (the "post-seed groups"), each comprised of at least one expert from each seed group so all students will learn about each energy source (see Fig. 26.1). This can be done by assigning students in each seed group a number. For example, in seed groups of five students, assign each student a number of one to five. All the 1's join post-seed group 1, the 2's join group 2, etc.

   To make a jigsaw activity work ideally, the number of students in a seed group should match the number of seed groups (see Fig. 26.1). However, this activity can work in classrooms of many sizes. For a large class, teachers can consider adding readings about more of the energy sources listed in Murphy (2012) including offshore oil (p. 7), shale gas (p. 10), biomass electricity (p. 16), concentrated solar thermal (p. 19), hydrogen (p. 20), or micropower (p. 21). Then the seed (and post-seed) groups can have five or six members with a corresponding number of seed groups (e.g., six seed groups of six students). When the number of students in a class does not allow equal group sizes, more than one expert can be in the post-seed groups (e.g., seed group C might have an extra member, so one of the post-seed groups will have two students who learned about that seed group's energy sources).

tions about different energy sources, but Murphy (2012) is recommended for the student readings because it is more current.

- Jigsaw groups should be arranged following the instructions in Box 26.1 and Fig. 26.1. (For an article about the jigsaw technique, see "National Institute for Science Education," n.d.)

Before class, the instructor will need to print:

- A blank copy of the "Energy Source Table" handout for each student (ESM-B).
- A copy of two different energy source readings (from Murphy 2012) for every student as determined by the sources assigned to their seed group (see Box 26.1). In smaller classes, all energy sources may not be used.
- One "World Energy Use" graph per group (ESM-C).

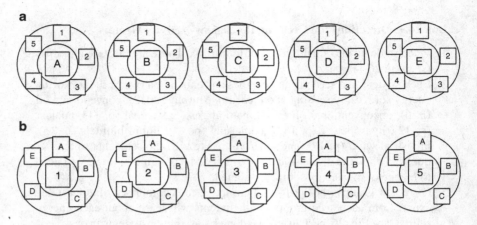

Fig. 26.1 Jigsaw seed and post-seed groups. (**a**) Seed groups (designated with letters A–E) from a class of 25 students. Each seed group is given two of the energy sources to read about (see Box 26.1), fill out the table for (ESM-B), and become "experts" on. Each student is given a number (1–5) to indicate their post-seed groups. (**b**) Post-seed groups, designated with numbers, for a class of 25 students. Each is comprised of one student from each of the lettered seed groups (e.g., post-seed group 1 was formed from students assigned a 1 in seed groups A, B, C, D, and E).

Activities

To start the activity, the instructor should explain that there are many considerations when looking at the possibility of using alternative energy sources instead of fossil fuels. The goal for the activity—to gain a better sense of the "landscape of energy choices" (sensu Murphy 2012) and to be able to describe the net energy ratio, the advantages, disadvantages, and externalities of energy sources—can be shared to help frame it for students. As desired, the instructor should also introduce the concepts of externality and net energy ratio.

Students should then be asked to join their assigned seed group and be given the handouts of the "World Energy Use" graph (ESM-C) and two energy sources (from Murphy 2012) that they need to read and a table (ESM-B) to complete (20 min). As the students complete their tables, the instructor should visit the groups to provide guidance and additional information as needed. For example, students may need direction in thinking about externalities; although the term "externality" does not only refer to disadvantages, students will most likely find that the externalities of their given energy sources will be negative.

Next, in a "jigsaw" manner, each seed group member moves to their assigned post-seed groups such that all energy sources are represented in these second groups. In the post-seed groups, each group member is considered an "expert" on his or her assigned energy sources and should share the findings about them from the chart (ESM-B) with the rest of the group (~5 min per student for a total of 20–30 min).

After the jigsaw group work, if time permits (or during the next class meeting) a class discussion can be fostered using questions to help students "dig a little deeper" into what they have learned about the different energy types (see ESM-D). If preferred, the instructor can give these questions to the students as homework or use them for assessment purposes.

Follow-Up Engagement

- A potential way to extend and facilitate the assessment of student learning is to have students write brief "minute papers" that respond to questions such as the following: What do you think about what you learned through this activity? What should individuals and governments do about energy issues? What can you do to educate others about the different types of energy?
- Students can be asked to write a journal entry that requires them to reflect on the obstacles to utilizing renewable energy and recommendations for potential implementation at a regional, state, and/or federal level.
- To help students understand how energy is personally relevant to them, an assignment could be given that asks them to investigate how their classroom building is heated/cooled, the major source(s) of electrical energy for their town, or the nearest location with certain types of energy generation facilities (such as a nuclear power plant or geothermal well) (see Chap. 24, this volume).
- To connect students' lives to energy use and pollution externalities, students can compute their own carbon footprint online (http://www.nature.org/greenliving/carboncalculator/). The website provides background information about carbon footprints, tips for offsetting one's carbon footprint, and other climate-related issues.

Connections

- Energy issues highlighted by this activity can be discussed in the context of environmental citizenship, focusing on what individuals can do to reduce energy use in their own lives and reasons that can be conveyed when encouraging others to reduce their energy use (see Chap. 6, this volume, for a related activity).
- This lesson relates to the triple-bottom-line concept of sustainability (which factors in ecological, economic, and social impacts; see Chaps. 2, 23, and 24, this volume). How do the types of energy that provide the most economic gain affect the economy and society? Is there a certain type of energy that seems to be the most sustainable from the triple-bottom-line perspective?
- For connections to the physics of energy, research projects about energy conversion (e.g., from sources such as wind into usable electricity) can be assigned (e.g., Chap. 25, this volume).

References

Heinberg R (2009) Searching for a miracle: "net energy" limits & the fate of industrial society. Post Carbon Institute & The International Forum on Globalization. http://www.postcarbon.org/publications/searching-for-a-miracle/. Accessed 18 Apr 2015

Murphy D (2012) The landscape of energy. In: Butler T, Wuerthner G, Lerch D (eds) The energy reader. Watershed Media, Healdsburg, CA, pp 105–152, http://energy-reality.org/wp-content/uploads/2013/01/14_The-Landscape-of-Energy_R2_122812.pdf

National Institute for Science Education (n.d.) Doing CL. http://www.wcer.wisc.edu/archive/cl1/cl/doingcl/jigsaw.htm. Accessed 14 Apr 2015

Part IV
Environmental Justice and Pollution

Chapter 27
Introducing the Conflicting Meanings of "Justice" Using a Candy-Distribution Exercise

Kate J. Darby

Introduction

Environmental justice (EJ), as both a body of scholarship and social movement, is concerned with the intersections of social justice and environmental issues. By the 1980s, scholars and activists began to document anecdotal evidence suggesting that locally undesirable land uses (LULUs), including landfills, toxic waste sites, and hazardous industrial facilities, were more likely to be located in low-income communities and communities of color than in wealthier, white neighborhoods (for historical accounts, see Cole and Foster 2000 and Gottlieb 2005; also see Chaps. 28–31, this volume). Key studies conducted by the United Church of Christ (1987) and Robert Bullard (1983) established empirical evidence for this pattern. In 1991, EJ activists organized a national conference (FNPCES 1991) at which they articulated the goals of the movement and President Clinton signed an executive order requiring federal agencies to consider EJ concerns, which led to the creation of an EJ program within the US EPA (EPA 2015). Since that time, activists and scholars have broadened their consideration of EJ issues to include many socio-ecological concerns related to power, privilege, and the environment, including disproportionate access to environmental amenities and resources (e.g., parks and open spaces, clean water and air), climate change issues (e.g., unequal impacts and tensions between economic development needs and reducing greenhouse gas emissions), and food and agriculture (e.g., farmworkers' health and rights, elitism in sustainable food movements).

Electronic supplementary materials: The online version of this chapter (doi:10.1007/978-3-319-28543-6_27) contains supplementary material, which is available to authorized users.

K.J. Darby (✉)
Huxley College of the Environment, Western Washington University, Bellingham, WA, USA
e-mail: kate.darby@wwu.edu

As awareness about EJ has grown, more instructors have integrated these topics into their courses. Some students enter environmental and sustainability studies courses with very strong ideas about injustice and passionate abhorrence for anyone with conflicting views, while others have not thought about justice in context of environmental concerns. The exercise described below was designed to challenge such views and introduce EJ by asking students to critically examine what constitutes a "fair" or "just" situation. This is accomplished through an activity in which students distribute resources (candy and hypothetical parks) and hazards (hypothetical industrial sites) among their classmates and then discuss both the outcomes and their decision-making processes. Ultimately, the exercise will demonstrate that "justice" is not a straightforward concept (Schlosberg 1999; Shrader-Frechette 2002) and that decisions that seem just for one person or group might be unjust for another.

Learning Outcomes

After completing this activity, students should be able to:

- Compare and contrast different ideas of justice.
- Apply ideas of justice to environmental controversies.
- Develop their own personal definition of justice in the context of environmental issues.

Course Context

- Designed for lower- and upper-level courses with 10–40 students, including environmental justice, history, and ethics courses
- 35 to 60 min in one class meeting
- No student background knowledge or preparation needed
- Adaptable to sustainability and environmental studies courses of any class size (if class is broken into smaller groups)

Instructor Preparation and Materials

This activity works well as an introduction to concepts related to environmental justice; if the course includes EJ content, this exercise should be used before introducing other EJ materials. If the instructor has a limited background in environmental justice, she/he may want to begin by familiarizing themselves with concepts from the field. The table in Electronic Supplementary Materials (ESMs) A summarizes some core EJ terms and questions to guide introductory content and discussions with students (that could be used to prepare handouts and presentation slides as desired). In addition, the introductory section of the Toxic Waste at Twenty report (Bullard et al. 2007) and Holifield's piece (2001) on defining environmental justice provide useful framing for this exercise. Instructors teaching natural resource issues might find the reader by Mutz et al. (2002) to be helpful. (Also see Chaps. 28–31, this volume, for additional views and resources.)

The instructor should prepare two sets of cards or slips of paper depicting environmental amenities and disamenities (e.g., ESM-B). For the first set of cards, find an image depicting a smokestack or other environmental disamenities (hazard) and print enough copies for one-half to three-quarters of the students. For the second set of cards, find an image depicting a park or other environmental amenities, and print out enough copies for one-quarter to one-half of the students. The number of environmental disamenities (hazards) and amenities should not be equal. The instructor should consider finding images that will be meaningful to the students, perhaps reflecting nearby locations and issues. The instructor should also acquire a bag of individually wrapped (as the students will be handling them) candy pieces that includes more pieces than there are students in the class. This is an important consideration. If presented with the same number of candy pieces as students in the class, students may be more likely to distribute the candy equally. This makes the resulting discussion less interesting and productive. Similarly, in the author's experience, providing fewer candy pieces than there are students in the class leads some students to quickly offer not to have any candy, and the exercise is completed without a great deal of deliberation; this also limits the ability to facilitate a productive class discussion about the exercise.

Activities

The following steps guide students through a set of decision-making processes to allocate environmental amenities (e.g., candy pieces and parks) and disamenities (e.g., smokestack). The instructor will provide some prompts, but the decision-making processes and outcomes will be completely executed by the students:

1. Explain to students that this activity should facilitate their thinking about environmental justice, which refers to the patterns of the distribution of environmental harms and benefits, and the decision-making processes around these issues. Encourage them to not overthink the exercise while participating, and explain that there will be a broader discussion about the issues it raises after completing it (2 min).
2. Place a bag of candy in the middle of the classroom and ask students to distribute it among themselves as they see fit. Tell them that you will not participate, that they must distribute ALL of the candy, and that they should let you know when they are finished. After all of the candy has been distributed, ask them to reflect on how and why they shared resources as they did and if this was a difficult task. State that the next task will likely be more difficult, as it involves difficult considerations of trade-offs (5–15 min).

3. Place the amenity and disamenity cards in the middle of the classroom. Ask the students again to distribute these amenities and disamenities; all cards must be given out and students may acquire more than one card. In this exercise, having the card means that a student lives within one-fourth mile of the image on the card. Again, tell them that you will not participate and that they should let you know when they are finished (5–10 min).

4. After students distribute all the cards, ask them to reflect on the outcome and process of each activity (the candy activity and the amenity/disamenity activity) in a short guided writing response. What was the outcome of each activity and to what extent is that outcome fair or just? What process did the class use to distribute the candy, environmental amenities, and disamenities? To what extent were those processes fair or just? (5 min).

5. Class reflection and discussion can be facilitated using the starting question "How did you distribute the candy and cards and why?" and those in ESM-A. In larger classes, instructors may wish to have students discuss these questions in small groups and reconvene to discuss common and divergent responses. The instructor should use the guiding questions to draw out and illustrate the different approaches to justice described in ESM-A and use the board/screen to underscore the new concepts and terms introduced (10–15 min).

6. Ask students to share with their neighbor how their ideas about fairness and justice have or have not changed as a result of the activities and discussion. The instructor should ask the students to summarize what they have learned through these exercises and then record these insights on the board. At this point, the instructor may wish to introduce the key distributive environmental injustices in the USA; in most cases, LULUs (locally undesirable land uses, such as toxic release inventory sites and landfills) tend to be located in communities of color and low-income residents (Bullard et al. 2007, Chap. 3). Explain that environmental justice activism, which tends to focus on concerns about the "environments" where people "live, work, and play," emerged in the 1980s as a way to address environmental issues not being addressed by the mainstream environmental movement (Gottlieb 2005) (5–10 min).

Follow-Up Engagement

- The Executive Summary and Chap. 1 from the United Church of Christ's "Toxic Waste at Twenty" report (Bullard et al. 2007) provide a nice set of follow-up readings to help students understand the historical development of environmental justice as movement and scholarship.
- Ask students to conduct research on the unjust distribution of LULUs and identify two or three procedural or structural factors (see ESM-A for definitions) that led to the unjust outcomes.
- Share a news report about a recent environmental controversy, or have students find their own examples. Ask them to explore the ways different forms of justice

are supported or denied by various actors in the environmental controversy case study.
- Have the students form groups and ask each group to find an organization working on environmental justice. Ask them to identify the advocacy approaches used by this organization. Students could also be invited to engage in some form of environmental justice activism for a cause that they personally support and reflect on their engagement in a follow-up essay.
- Relevant activities are described in Chaps. 28–32 of this volume.

Connections

- Issues of environmental justice can be related to energy extraction (e.g., mountaintop removal coal mining; Chaps. 25 and 26, this volume), food and agriculture (e.g., farmworkers' rights and pesticide drift; Chap. 21), climate change (e.g., implications for small island nations and the idea of "just" sustainable economic transitions; Chap. 39), and urban planning (e.g., urban redevelopment/gentrification, providing access to green spaces designed with the community's needs in mind; Chaps. 19 and 20).
- Discussions of current environmental controversies can be linked to the underlying conceptions of justice and fairness valued by various stakeholders. For example, recent controversies surrounding hydraulic fracturing for natural gas extraction can be framed and examined with the following questions:

    - From an economic perspective, who benefits from these activities and who is burdened?
    - What are the environmental impacts from hydrofracking, especially those related to human health? Who is most affected by these impacts?
    - Who makes decisions about hydraulic fracturing and how are those decisions made?
    - What are the long-term impacts of hydraulic fracturing on communities?

References

Bullard RD (1983) Solid waste sites and the black Houston community. Socio Inq 53:273–288
Bullard RD, Mohai P, Saha R, Wright B (2007) Toxic wastes and race at twenty (1987–2007). United Church of Christ. http://www.ejnet.org/ej/twart.pdf. Accessed 21 Aug 2015
Cole L, Foster S (2000) From the ground up: environmental racism and the rise of the environmental justice movement. NYU Press, New York
First National People of Color Environmental Summit (1991) Principles of environmental justice. http://www.ejnet.org/ej/principles.html. Accessed 21 Aug 2015
Gottlieb R (2005) Forcing the spring: the transformation of the American environmental movement. Island Press, Washington, DC

Holifield R (2001) Defining environmental justice and environmental racism. Urban Geog 22:78–90

Mutz K, Bryner G, Kenney D (2002) Justice and natural resources: concepts, strategies, and applications. Island Press, Washington, DC

Schlosberg D (1999) Environmental justice and the new pluralism: the challenge of difference for environmentalism. Oxford University Press, Oxford

Shrader-Frechette KS (2002) Environmental justice: creating equality, reclaiming democracy. Oxford University Press, Oxford

United Church of Christ (1987) Toxic wastes and race in the United States: a national report on the racial and socio-economic characteristics of communities with hazardous waste sites: Public data access, Inc. www.ucc.org/about-us/archives/pdfs/toxwrace87.pdf. Accessed 21 Aug 2015

(EPA) U.S. Environmental Protection Agency (2015) Environmental justice. http://www.epa.gov/environmentaljustice/. Accessed 21 Aug 2015

Chapter 28
Beyond Band-Aids: Using Systems Thinking to Assess Environmental Justice

Meghann Jarchow

Introduction

Environmental justice refers to the fair distribution of environmental benefits and burdens across all human populations (Agyeman 2005; also see Chaps. 27, 29 and 30 of this volume). Ameliorating environmental injustices, which often disproportionately impact low-income and minority communities (Bullard 2000), requires using a systems-thinking approach which emphasizes the importance of understanding how complex interactions and relationships affect the functioning of systems (Meadows 2008; Chaps. 3 and 31 of this volume). This applies to addressing environmental injustices because they emerge from complex relationships among environmental, social, and economic systems, and systems thinking provides a framework for elucidating ways to intervene in a system. In contrast, more simplistic "Band-Aid" interventions, which may ameliorate an injustice in the short term or at a local level, are unlikely to create long-term, systemic change and therefore allow the root causes of injustices to persist. Band-Aid interventions fail to get to the "heart of the problem"—like taking aspirin to cover up pain without identifying the cause of the pain. Interventions to root causes of injustices are more likely to create long-term, systemic change, but root causes are often more difficult to identify and change.

Environmental justice and systems thinking are important topics in environmental and sustainability studies. The goal of this learning activity is to help students

Electronic supplementary materials: The online version of this chapter (doi:10.1007/978-3-319-28543-6_28) contains supplementary material, which is available to authorized users.

M. Jarchow (✉)
Sustainability Program, University of South Dakota, Vermillion, SD, USA
e-mail: meghann.jarchow@usd.edu

use a systems-thinking approach to examine and then propose systemic interventions for environmental injustices. To accomplish this, students watch one of six short videos addressing environmental justice issues outside of class and complete an accompanying worksheet. In class, groups of students who watched the same video evaluate the issues using a systems-thinking approach, propose interventions to alleviate the injustice, and present their findings to the class.

Learning Outcomes

After completing this activity, students should be able to:

- Describe environmental injustices affecting disadvantaged populations.
- Recognize how systems thinking applies to examining environmental (in)justice.
- Differentiate between Band-Aid and systemic interventions to alleviate environmental injustices.
- Propose Band-Aid and systemic interventions to alleviate an environmental injustice.

Course Context

- Developed for an introductory sustainability course with ~30 students that was required for a major and minor sustainability but also met a social science general education requirement
- 50 min in one class meeting
- Before class, students watch a short video and complete an accompanying worksheet, but no additional background knowledge is needed
- Adaptable to a wide range of courses and class sizes; can be modified for alternative class sizes by altering the number and size of groups and/or the number of videos watched

Instructor Preparation and Materials

Instructors should be familiar with systems thinking (Meadows 2008) and environmental justice and injustice (Schlosberg 2004; Agyeman 2005; Brulle and Pellow 2006) and be able to discuss how they are related (also see Chaps. 2, 27, 29–31 for more perspectives and resources). The recommended environmental justice videos (Box 28.1) should be viewed and decisions made about which ones to assign to students prior to the activity. In the author's experience, each chosen video should be assigned to four to six students, which then form the in-class groups. Instructors should review—and edit as needed—the slide presentation (Electronic Supplementary Material (ESM) A), which can be used during class, and the root-causes handout (ESM-B) and student worksheet (ESM-C), which can be assigned as before-class preparation materials for the students to complete as they watch the videos. As an example of the types of answers expected on the student worksheet, ESM-D contains suggested answers for one of the videos. Additional videos are suggested in ESM-E.

Activities

Before the class in which this activity is implemented, students should be divided into groups and assigned an environmental justice video (i.e., each group will watch a different video; Box 28.1). After watching the video, students should complete the worksheet to prepare for the in-class activity (ESM-C).

During class, the instructor should introduce the concepts of systems, systems thinking, Band-Aid and systemic interventions, and root causes (5 min; ESM-A). After this, students should join their groups and discuss the completed worksheet questions (~10 min). For an additional 10 min, each group should then identify two interventions, one Band-Aid and one systemic, to ameliorate the environmental injustice(s) described in their video.

Box 28.1. Environmental Justice Videos (See ESM-E for Website Links and Additional Descriptions)

- "Ghana: Digital Dumping Ground"—This 20-min video explores the social and environmental impacts of e-waste (electronic waste) in Ghana, China, and India. (See ESM-D for answers to worksheet questions and a concept map for this video.)
- "Majora Carter: Greening the Ghetto"—In this 19-min talk, Majora Carter describes the disproportionately high and low amounts of industry and green spaces, respectively, in the South Bronx where she lives, and community-led revitalization efforts.
- "Barry Estabrook: Following the Money in Lake Apopka, FL"—In this 13-min presentation, Barry Estabrook describes the history of Lake Apopka, FL, especially the history of the wealthy landowners who grew fruit and vegetable crops there and the largely African American farm workers who managed and picked the crops and have been negatively affected by pesticides.

(continued)

Box 28.1. (continued)

- "Van Jones: The Economic Injustice of Plastic"—In this 13-min talk, Van Jones argues that our attitude of disposability is an environmental justice issue because it disproportionately affects poor countries and poor people.
- "U. Roberto Romano: Children in Our Fields"—In this 12-min presentation, U. Roberto Romano describes how a large percentage of the hand-picked food that we eat in the United States is picked by children migrant farm workers who are US citizens but are largely of Mexican descent.
- "Peggy Shepard: Environmental Justice"—In this 8-min talk, Peggy Shepard, one of the founders of the idea of environmental racism, describes "sacrifice zones," which are areas that have disproportionately high environmental damage or economic disinvestment.

Each group should then present their work to the class by having one member of the group briefly summarize (for a duration determined by the number of groups and class period length) their group's environmental justice issue and describe the systemic intervention that the group proposed. The instructor should remind students that the rest of the class did not watch their video and therefore did not learn about the same environmental justice issue. As the groups are presenting, the instructor should take notes on the board to facilitate summary and discussion. Following the presentations, the instructor should summarize the presentations and highlight the range of disadvantaged populations susceptible to environmental injustices and the similarities or differences (as applicable) in root causes identified by groups. Students may think that Band-Aid interventions are less desirable or ineffective, but the instructor should suggest that Band-Aid interventions are often necessary and important because they generally have more immediate effects and are easier to change. As desired and time permits, additional discussion and student reflection could be fostered to help students achieve the learning outcomes. For example, the instructor could provide an example of an initiative being led in the local community. The class could discuss whether it is a Band-Aid or systemic intervention and why that intervention was likely chosen.

Follow-Up Engagement

- At the end of the class period, give students a case study of another environmental justice issue (e.g., one affecting a different population such as Native Americans), and ask them to write a brief essay addressing the worksheet questions (ESM-C).
- In a subsequent class period, have the students use the content from all of the groups' work to create a concept map about environmental justice (see ESM-D for an example; also see Chap. 31 of this volume).

- Ask the students if they, or someone they know, are part of one of the disadvantaged groups and have experienced environmental injustices. If yes, ask them to describe those injustices. For their examples or more generally, ask them what roles they might have in promoting environmental justice.
- Related activities are described in Chaps. 27, 29, and 30 of this volume.

Connections

- If root causes (ESM-A) are discussed in additional class periods, have students reflect on how the same root cause can underlie seemingly disparate issues such as the example provided in the presentation.
- When examining environmental changes such as climate change, habitat loss, and water pollution, ask students if there are any environmental injustices that are also happening (e.g., relating to activities in Chaps. 13, 15, 16, 19, and 20).
- Have student apply a systems-thinking approach to analyze another complex social systems topics such as an environmental policy or law or the activities and environmental impacts of an international company (e.g., Chaps. 3 and 22–24 of this volume).

References

Agyeman J (2005) Sustainable communities and the challenge of environmental justice. NYU Press, New York

Brulle RJ, Pellow DN (2006) Environmental justice: human health and environmental inequalities. Annu Rev Publ Health 27:103–124

Bullard RD (2000) Dumping in Dixie: race, class, and environmental quality, 3rd edn. Westview Press, New York

Meadows DH (2008) Thinking in systems: a primer. Chelsea Green Publishing, White River Junction

Schlosberg D (2004) Reconceiving environmental justice: global movements and political theories. Environ Polit 13:517–540

Chapter 29
Engaging the Empathic Imagination to Explore Environmental Justice

Joseph Witt

Introduction

Environmental justice refers to the "fair treatment and meaningful involvement of all people regardless of race, color, national origin, or income with respect to the development, implementation, and enforcement of environmental laws, regulations, and policies" (EPA 2015; also see Chaps. 27, 28 and 30 of this volume). The vision of "justice" in this framework is most often conceived as distributive justice, where "just" activities are those where costs and benefits are fairly and equitably distributed, while unjust activities are those where one individual or group receives an unequal burden of cost, while another group reaps an unequal and unfair set of benefits and rewards (see Rawls 1971). Research has shown that, frequently, environmental problems are unequally distributed along sociodemographic lines such as race, gender, and class. Environmental justice movements seek to correct these related social injustices along with resolving connected environmental issues (Bullard 2000; Gottlieb 2005).

The activity described here is designed to help students empathetically encounter individuals experiencing different levels of environmental injustice and "go out toward people to inhabit their worlds, not just by rational calculations but also in imagination, feeling, and expression" (Johnson 1993: 200). By considering the stories of individuals directly facing environmental challenges, students attempt (as much as possible) to imagine themselves in the embodied, lived experiences of persons different from themselves to arrive at new, creative

Electronic supplementary materials: The online version of this chapter (doi:10.1007/978-3-319-28543-6_29) contains supplementary material, which is available to authorized users.

J. Witt (✉)
Department of Philosophy and Religion, Mississippi State University, Starkville, MS, USA
e-mail: jwitt@philrel.msstate.edu

© Springer International Publishing Switzerland 2016
L.B. Byrne (ed.), *Learner-Centered Teaching Activities for Environmental and Sustainability Studies*, DOI 10.1007/978-3-319-28543-6_29

perspectives on environmental justice and inequality. The goal is to engage students' empathetic imagination to step out of their own assumptions, lived experiences, and worldviews and imagine new perspectives, narratives, and experiences surrounding issues of environmental justice. Classroom activities such as these that engage students' empathetic imaginations may even help motivate greater active participation in projects promoting sustainability and environmental justice, as suggested by recent psychological research (Pfattheicher et al. 2015; also see Chaps. 4, 19, 39, 41, this volume).

Learning Outcomes

After completing this activity, students should be able to:

- Describe an environmental issue from a perspective different from their own.
- Analyze the influences of sociodemographic factors such as race, gender, class, and ability (among others) on the experiences of environmental harms and benefits among diverse individuals.
- Assess the value of empathizing with the experiences of others to develop solutions to problems of environmental injustice.

Course Context

- Developed for an upper-level course on religions and the environment for both majors and nonmajors with 25–35 students
- 75 min in one class meeting
- No previous knowledge of environmental justice is needed for students to perform this exercise, but instructors may choose to introduce themes and concepts related to environmental justice in class sessions before this exercise
- Adaptable to classes of various sizes focused on environmental humanities and social inequalities; may be adapted to larger class sizes by dividing students into small groups

Instructor Preparation and Materials

Instructors should prepare two resources to conduct this exercise: (1) a short narrative of an environmental issue to be considered by all students and (2) a set of individual biographies that students will read and use to take the perspective of another person in order to evaluate the given issue. For the first component, the chosen environmental issue may be a fictional case, but it should be based in real events (e.g., the case of a new oil pipeline to be built through a region of the US Midwest has been successfully used by the author; see Electronic Supplementary Materials (ESM) A for this example and ESM-B for other issues). The narrative of the issue should briefly describe basic details about its environmental and social costs and benefits and point to possible areas of debate surrounding it. It should give students enough information to develop their own responses and opinions. Along with the narrative, instructors should prepare a list of questions to include on the biography handouts that will guide students' reflections. Students will answer these questions from the perspective of their given biography; suggestions are provided in ESM-A.

Second, instructors must craft a set of individual biographies of fictional characters (less than a page each) who will be impacted by the issue in some way. The biographies should represent a diverse set of backgrounds, sociodemographic characteristics (e.g., different races, genders, and levels of wealth and education), and experiences, each with some connection to the chosen issue. It is important that these accounts, though fictional, should be realistic, offer no clear "good" or "bad" characters, and avoid stereotypes. (Example biographies for the pipeline construction case study are provided in ESM-A.) Each student will receive only one biography, and it is recommended that instructors draft about five distinct biographies to reflect a diversity of perspectives. Distribution of the biographies is dependent on class size. In classes with smaller enrollment, instructors may distribute the five biographies randomly (with several students receiving the same biography). Students will share their biographies with the whole class during the second and third phases of the exercise (see Activities). For classes with larger enrollment, instructors may separate students into small groups with each student in the group receiving one of the different biographies; in this context, phase two of the exercise can be conducted in small groups which then report back to the full class as a group in phase three.

A biography should be distributed to each student as a printed handout. If small groups are used, the number of group members should be chosen so that each group receives a complete set of different biographies (as needed, multiple students with the same biography is acceptable). In class, students will need writing materials to record their thoughts as they read and evaluate the case, and the teacher will need access to a white-/chalkboard or other presentation materials to record themes from students' responses during the group discussion. The narrative of the issue under consideration and the reflection questions should also be made available to students, either on the individual handouts or projected onto a screen throughout the duration of the exercise.

Before using this exercise, instructors should also decide whether they want to introduce concepts and theories surrounding environmental justice beforehand or perform this exercise "cold," waiting to introduce core concepts until after the exercise, thus allowing students to form conclusions without influence from others' perspectives. Their responses to this exercise could then be used as data to explore perceptions of environmental justice and the value and possibility of empathy in later class sessions. Either way, themes and concepts of environmental justice could be introduced through discussions based on introductory readings, including Robert

Bullard's introduction to *The Quest for Environmental Justice* (2005) or selections from Gordon Walker's *Environmental Justice* (2012).

Activities

This exercise progresses through three phases. First, the instructor should distribute the individual biographies to the students and explain the instructions (10 min). Read the description of the environmental issue together as a group and then explain that students should individually read their given biography and answer the reflection questions (ESM-A) *from the perspective of their given character's biography* (10–15 min). (Whether they should formally write down their reflections is at the discretion of the instructor.) If students have questions about their given biography that are not included on the handout (e.g., the person's age, name, place of birth, etc.), they should be encouraged to fill in those details on their own to enable them, as much as possible, to take on the perspective of the fictional person and feel a stronger empathetic connection which can increase the impact of the exercise.

In the second phase, invite students to share and discuss the details of their biographies and responses with the class (or their group members if groups are used) (15–25 min). For this phase, students should be speaking *as their given characters* (i.e., role play). How did they answer the questions from the perspective of their given biography? Until this point, students will have been unaware of the other biographies or who else in the class (or group) was given their same scenario, and they should be free to ask questions of each other and explore differences among the different characters' responses. Instructors should keep track of major themes on a white/chalkboard as they emerge. Based on the author's experiences using the pipeline case study (ESM-A), students most generally noted issues such as age, gender, religion, education, wealth, and political agency as influencing characters' experiences of the issue, but any of a wide array of factors may be evident depending upon how the biographies are prepared.

For the third phase of the exercise (15+ min), students should return to *their own viewpoints* (not those of their given biographies) to examine the major themes of the exercise. Having listed relevant factors influencing characters' experiences of the issue from the second phase, students may now be asked to reflect on and share the challenges they encountered in taking new perspectives or empathizing with the embodied experiences of another. Some pointed questions to guide discussion include: How did the perspective of your given biography (as presented by you) differ from your own personal opinions about this issue? Is it possible to truly empathize with another? What is the value of taking on perspectives of others? How might this influence your own opinion about environmental issues? In the author's experience, students who had the same biography often arrived at slightly different evaluations of the issue. For example, in a recent usage of this exercise with the pipeline case study, one student with the biography of the American Indian woman reported feeling opposed to the pipeline because of its impacts on her culturally sacred lands; another student with the same biography, however, reported that she would be happy that the pipeline would bring new jobs to her community. This and other differences between student accounts led to very lively discussion surrounding the limitations of truly empathizing with others and setting aside one's own ideological perspectives. Through the discussion, instructors should help students come to appreciate that different people experience environmental issues differently based on various sociodemographic categories and that such differences influence people's views and decisions about issues. As such, discerning "just" solutions to environmental problems requires empathizing with the perspectives of others and taking these differences seriously.

Follow-Up Engagement

- Students may be asked to reflect further on the exercise and relate it to their own personal experiences through discussion questions such as:
  - What, if anything, was especially difficult or challenging about this exercise?
  - Do people empathize with others enough in public debates? Why or why not?
  - Do you think empathizing with the experiences of others may help to motivate actions challenging environmental injustices or not?
- With the assignment, students reflect on the issue immediately after reading their fictional biographies. They could also be asked to return to the issue several days later, perhaps after engaging with more scholarly sources on environmental justice. They can then be asked to compare their initial responses with those several days later, pointing to how their thoughts on environmental justice have changed over time.
- Bring in films or news reports about recent environmental events and ask students to analyze those using their empathetic imaginations (e.g., videos provided in Chap. 28 of this volume).
- Invite students to have a formal, but friendly, debate on the specific issue (e.g., as in a town hall meeting), arguing from the perspectives of their given characters (e.g., Chap. 39 of this volume).
- Additional related activities are described in Chaps. 27 and 30 of this volume.

Connections

- When other topics are introduced, invite students to identify environmental justice issues associated with them and then imagine the experiences and perspectives of different stakeholders.

- Discuss what other psychological and social factors (e.g., religion, worldviews, ethics) might influence environmental perspectives and how those influence our understanding of environmental justice and, in turn, individual and collective decision-making about environmental issues. (See related topics and activities in Chaps. 4, 6, 8, 19, 21, 22, and 24 of this volume.)

References

Bullard RD (2000) Dumping in Dixie: race, class, and environmental quality, 3rd edn. Westview Press, Boulder

Bullard RD (ed) (2005) The quest for environmental justice: human rights and the politics of pollution. Sierra Club Books, San Francisco

EPA (2015) What is environmental justice? http://www.epa.gov/environmentaljustice/. Accessed 4 July 2015

Gottlieb R (2005) Forcing the spring: the transformation of the American environmental movement. Island Press, Washington, DC

Johnson M (1993) Moral imagination: implications of cognitive science for ethics. University of Chicago Press, Chicago

Pfattheicher S, Sassenrath C, Schindler S (2015) Feelings for the suffering of others and the environment: compassion fosters proenvironmental tendencies. Environ Behav. doi:10.1177/001391 6515574549

Rawls J (1971) A theory of justice. Harvard University Press, Cambridge

Walker G (2012) Environmental justice: concepts, evidence, and politics. Routledge, New York

Chapter 30
Helping Students Envision Justice in the Sustainable City

Joshua Long

Introduction

In North America, where more than 80 % of the population is urbanized, cities are increasingly taking a leadership role in promoting sustainability (Benton-Short and Short 2013). However, no agreed-upon definition of a "sustainable city" exists (Lorr 2012). Instead, city leaders and policymakers frequently employ the "three pillars" approach to defining goals according to economic, social, and environmental concerns (e.g., see Chap. 2, this volume). Creating a sustainable city relies upon a careful balance of all three pillars with all residents in mind. Yet, as cities advance projects intended to reduce their eco-footprint, improve the health and well-being of their citizens, and sustain their local economies, they face the challenge of just and equitable implementation (Dempsey et al. 2009; Benton-Short and Short 2013; also see Chaps. 27–29, this volume). The development of mass transit, the creation of a city park or greenbelt, and the implementation of a new waste disposal program are initiatives that can often create economic, social, and environmental trade-offs that are realized disproportionately among residents. Even those cities with excellent reputations for sustainability endure significant disparities and inequalities that undermine the overall goals of sustainable development (Schweitzer and Stephenson 2007; Wolch et al. 2014). As Benton-Short and Short (2013, p. 481) note: "Moving toward sustainability is not simply 'greening' the city… a sustainable city is also a just city."

Many of the social aspects of urban sustainability have been largely underrepresented in sustainability initiatives. The goal of this chapter's learning activity is to reveal the challenges and complexities of equitable planning for urban sustainability

Electronic supplementary materials: The online version of this chapter (doi:10.1007/978-3-319-28543-6_30) contains supplementary material, which is available to authorized users.

J. Long (✉)
Environmental Studies, Southwestern University, Georgetown, TX, USA
e-mail: jlong@southwestern.edu

© Springer International Publishing Switzerland 2016
L.B. Byrne (ed.), *Learner-Centered Teaching Activities for Environmental and Sustainability Studies*, DOI 10.1007/978-3-319-28543-6_30

and to develop students' critical thinking skills that can help them identify the economic, social, and environmental disparities within urban landscapes. By comparing the lives of two urban citizens from different parts of the city, students will explore how each of these citizens gains access to services and amenities that are seen as part of living in a sustainable city. The exercise ends by asking students to consider ways of promoting justice and equitable development in the sustainable city.

Learning Outcomes

After completing this activity, students should be able to:

- Critically engage with the concept of "urban sustainability" to reveal the challenges of equitable sustainable development in a city.
- Describe the concepts of environmental and social justice as they relate to patterns of unequal distribution of social services, infrastructure, and environmental amenities in a city.
- Articulate their responsibilities as urban citizens pursuing a sustainable lifestyle and consider how their position may provide them with advantages or disadvantages compared to others in this pursuit.

Course Context

- Developed for an upper-level seminar on Urban Sustainability with 15 students
- 75 min in one class meeting
- Adaptable for larger or smaller courses that cover urban sustainability and environmental justice

Instructor Preparation and Materials

For general context of the activity, instructors should be familiar with the connections and tensions among economic vitality, social justice, and environmental protection in planning for urban sustainability (see Krueger and Gibbs 2007; Dempsey et al. 2009; Lorr 2012; Benton-Short and Short 2013). The instructor should understand the links between social and environmental justice and sustainable development and ways to best communicate this from a critical urban studies perspective (see Esposito and Swain 2009; Dempsey et al. 2009). For example, Schweitzer and Stephenson (2007) provide an excellent introduction to environmental justice (a phrase that incorporates a social justice dimension to "environmental racism") and its importance in urban policy. Wolch et al. (2014) offer important insights into the links between environmental sustainability initiatives and gentrification/displacement. Manaugh et al. (2015) broadly discuss the lack of planning for social equity in transportation issues in American cities. Since sustainable development is fraught with contradictions, the instructor may find it helpful to review case studies that reveal the challenges inherent to urban sustainability projects, particularly those that explicitly address social and environmental justice. For example, Goodling et al. (2015) uses the case study of Portland, Oregon, to reveal examples of the challenges of equitable implementation of sustainability policies and initiatives in one of America's most sustainable cities. Dooling (2009) describes how the ecological rehabilitation of an urban watershed in Seattle led to the displacement of the homeless and related social services. Each of the sources suggested above carefully juxtaposes the intended benefits of sustainability projects against the potential drawbacks, reminding the reader that both planners and the public have a responsibility to ensure that social and environmental justice remain key components of urban sustainability.

   The above resources can be used to construct an introductory lecture, create worksheets ahead of class that would facilitate and/or shorten the exercise, prepare definitions for terms or concepts that may be unclear, and/or provide examples to help students understand how these problems materialize in real cities. Familiarity with these sources may also provide instructors with the knowledge needed to tailor this exercise to individual classes. The discussion questions in Parts II and III below are provided in Electronic Supplementary Materials (ESM) A for instructors to use to prepare presentation slides, worksheets, or other materials.

Activities

Part I (20 min)

At the beginning of class, the instructor asks the students to imagine the ideal sustainable urban landscape and consider it in the context of the larger city. (As needed, also introduce the "three pillars" of sustainability to frame the discussions so that students recognize the need to consider economic, social, and environmental aspects of urban sustainability.) Thinking about the following features at the neighborhood scale will help to focus the exercise: transportation infrastructure, water quality, pollution, parks and greenbelts, economy and industry, social services, diversity, education, crime, entertainment and leisure activities, etc. Students are then asked to imagine a stereotypical *un*sustainable urban neighborhood, considering the same features as above. Students can be asked to write down brief notes that describe each of these cities, and the instructor may choose to write these on the board. Following this, the instructor then asks the class to collectively (or in small groups) construct a profile of an "average" resident of each city, recording the basic description of each to facilitate comparing and contrasting their lives. Creativity should be encouraged to make this part more relatable; for example, it is suggested that the class develop names, occupations, background stories, etc. for each of the residents. This exercise will obviously yield different results depending on the class, but for an example see ESM-B. Alternatively, the instructor can save time by using the supplementary example to facilitate the exercise.

Part II (30 min)

The instructor reminds the students that their imagined sustainable and unsustainable urban landscapes—and the lives of their representative citizens—exist in virtually every major city in North America, including those with excellent reputations for urban sustainability. For purposes of the exercise, the instructor then asks students to imagine that these two citizens live in the same city and that their lives are connected and interdependent economically, environmentally, and socially. In this part of the exercise, the instructor asks a series of questions intended to show the disparities and connections between these two citizens:

- (Economic) Is a sustainable lifestyle more expensive to maintain? For example, does it cost more to live in a "green" neighborhood, purchase sustainable food, or have access to public transportation? From a social sustainability perspective, would a higher salary allow access to better education, health care, entertainment, or other lifestyle amenities?
- (Economic) How are these lives connected or disconnected through their occupations, taxes and social services, and/or class structure? Which of these residents can respond more easily to increases in property taxes, rent, or costs of services?
- (Social) Which of these citizens has access to better educational opportunities and health care? Which is more likely to live in a high-crime neighborhood? How might access to entertainment and leisure activities differ between these two citizens?
- (Social) Which of these citizens is more likely to encounter some form of discrimination or displacement? When you were constructing the profiles of each of these citizens, did you picture them to have different ages, genders, levels of education, ethnic backgrounds, and family situations?
- (Environmental) Which is more likely to reside in a neighborhood with cleaner air, water, and soil? How might such environmental conditions impact their health and livelihood?
- (Environmental) Which of these citizens has greater access to parks and which lives closer to their work or has better access to sustainable transportation? Do both residents have access to the same recycling options, compost pickup, or other waste services?
- (Environmental) Which of these citizens has a larger eco-footprint and why?
- (Synthesis) How do these questions reveal the interdependency of sustainability's "three pillars" and the need to incorporate all three equitably to ensure true sustainability?

- (Synthesis) Refer to the list of features in Part I (i.e., transportation infrastructure, water quality, pollution, parks, etc.). How would each of these require consideration of social, economic, and environmental issues to address their sustainability?

Part III (25 min)

The instructor then asks the class to try to construct a definition of "urban sustainability." Following this, the instructor should ask the following:

- When considering the lives of the two citizens profiled in the earlier part of this activity, do you think that there may be different definitions of sustainability for different citizens? What might these look like? How could they all contribute to a city's sustainability?
- Did you find yourself relating to one of the two citizen profiles created in the first part of the activity? How does their situation affect their ability to live a more sustainable life? If you lived in the city, what could you do as an individual to live a sustainable life? How does your position as a college-educated person allow you certain advantages in the pursuit of sustainability? What other advantages might you have besides education?
- What projects or initiatives would you suggest that would make the city more sustainable in an equitable manner for the greatest number of citizens? In other words, if you are going to build a sustainable city, how do you make it sustainable for everyone and not just a privileged few?

Follow-Up Engagement

- Assign homework that requires students to develop a new sustainability initiative (such as constructing a new mass transit route, rehabilitating an urban watershed, or building a new waste-to-energy incineration plant). Students should consider who benefits from this project and who is negatively affected. They must employ all three "pillars" of sustainability and consider the trade-offs for different communities within the same city.
- Assign a group project in which some groups profile a historically low-income neighborhood of the city, while others profile an upper middle-class neighborhood. Each group must find data on property values, average income, ethnic makeup, educational attainment, history of hazardous waste or pollution, crime rate, access to transportation options, and additional factors chosen by the instructor. Each group presents on their neighborhood and discusses how policy measures (whether real or hypothetical) may affect each neighborhood differently.

- Ask students to summarize and respond to a recent newspaper article from a major US city that highlights how a specific sustainability initiative or environmental policy measure would affect residents in the city depending on their income, location, and/or other demographic characteristics. (For related skill-focused activities, see Chaps. 37 and 40, this volume.)
- Other chapters in this volume describe activities that can build on topics in this chapter including land use management and planning (Chaps.19 and 20), considering the worldviews of others (Chaps. 4, 39and 41), environmental justice (Chaps. 27–29), and pollution (Chaps. 31–33).

Connections

- Environmental and social justice issues can affect the overall eco-footprint of a city and therefore have connections to larger issues such as global climate change and biodiversity loss.
- Compare the environmental justice issues in the city to global environmental justice issues; how are they similar but on a much larger scale and how do they differ?
- How does urban sustainability compare to sustainability issues that affect rural areas and small towns?

References

Benton-Short L, Short JR (2013) Cities and nature, 2nd edn. Taylor and Francis, New York, NY

Dempsey N, Bramley G, Power S, Brown C (2009) The social dimension of sustainable development: defining urban social sustainability. Sustain Dev 19:289–300

Dooling S (2009) Ecological gentrification: a research agenda exploring justice in the city. Int J Urban Reg Res 33:621–639

Esposito J, Swain AN (2009) Pathways to social justice: urban teacher's uses of culturally relevant pedagogy as a conduit for teaching social justice. Persp Urban Ed. Spring: 38–48

Goodling E, Green J, McClintock N (2015) Uneven development of the sustainable city: shifting capital in Portland. Oreg Urban Geog. doi:10.1080/02723638.2015.1010791

Krueger R, Gibbs D (2007) The sustainable development paradox. Guilford, New York, NY

Lorr MJ (2012) Defining urban sustainability in the context of North American cities. Nat Cult 7:16–30

Manaugh K, Badami MG, El-Geneidy AM (2015) Integrating social equity into urban transportation planning: a critical evaluation of equity objectives and measures in transportation plans in North America. Transp Policy 37:167–176

Schweitzer L, Stephenson M (2007) Right answers, wrong questions: environmental justice as urban research. Urban Stud 44:319–337

Wolch J, Byrne J, Newell J (2014) Urban green space, public health, and environmental justice: the challenge of making cities 'just green enough'. Landsc Urban Plan 125:234–244

Chapter 31
Social-Ecological Systems Mapping to Enhance Students' Understanding of Community-Scale Conflicts Related to Industrial Pollution

Curt Dawe Gervich

Introduction

The federal Emergency Planning and Community Right-to-Know Act (EPCRA) (U.S. Code, Title 42, Chapter 116) was passed by the US government in 1986 after several national and international industrial accidents released toxic chemicals into the environment (EPA 2015a). These accidental releases, which include the 1984 Union Carbide disaster in Bhopal, India, carried vast ecological and public health consequences. Subsequently, elected officials and environmental regulators in the United States strongly supported legislation to document and raise awareness about the risks of industrial pollutants. EPCRA has two main purposes: (1) to assist state, county, and municipal governments in developing response plans for accidental chemical releases and (2) to provide residents with information about the toxic chemicals handled by industries in their communities. The steps put in place to raise awareness about industry's use of hazardous chemicals are outlined in the Toxics Release Inventory (TRI 1986). This program, which is overseen by the Environmental Protection Agency (EPA), includes strict reporting requirements for approximately 20,000 industrial facilities across the United States (EPA 2015a). The TRI empowers community groups with information to assist them in confronting industrial polluters and engaging government regulators to reduce local ecological and public health risks.

The activity described in this chapter provides an introduction to the TRI and highlights some of the ways that community groups in the United States use the

Electronic supplementary materials: The online version of this chapter (doi:10.1007/978-3-319-28543-6_31) contains supplementary material, which is available to authorized users.

C.D. Gervich (✉)
Center for Earth and Environmental Science, SUNY Plattsburgh, Plattsburgh, NY, USA
e-mail: Cgerv001@plattsburgh.edu

© Springer International Publishing Switzerland 2016
L.B. Byrne (ed.), *Learner-Centered Teaching Activities for Environmental and Sustainability Studies*, DOI 10.1007/978-3-319-28543-6_31

program. The activity uses a National Public Radio series, *Poisoned Places* (NPR 2011), as source material for exploring and creating social-ecological systems maps. The *Poisoned Places* case studies powerfully document local-level conflicts related to hazardous pollution. Social-ecological systems mapping is a tool for illustrating and analyzing the complex relationships among the human and more-than-human components of an environmental management or governance challenge. The main objective of this activity is to help students use social-ecological systems mapping to reveal the roles and responsibilities of industry actors, community organizations, and government regulators involved in the governance of toxic releases at community scales and to identify leverage points for de-escalating environmental conflicts among these groups.

Learning Outcomes

After completing this activity, students should be able to:

- Elaborate on the objectives, requirements, and uses of the Toxics Release Inventory.
- Create and describe a social-ecological systems map that illustrates the ways that industry, community, and government stakeholders interact to manage toxic pollutants.
- Identify the relationships among stakeholders that contribute to pollution-related conflicts at the community scale.
- Use social-ecological systems mapping to highlight leverage points for de-escalating conflicts and improving management of industrial pollutants.

Course Context

- Developed for an introductory-level environment and society course with 25–30 environmental studies and environmental science majors
- ~50 min of in-class time
- Students do not need preparation but may benefit from reading the EPA TRI Fact Sheets (provided as Electronic Supplementary Materials (ESM) A, B and C) or review the TRI website before the activity: http://www2.epa.gov/toxics-release -inventory-tri-program
- Adaptable for use in advanced environmental policy or environmental sociology courses. In policy-focused courses, instructors may want to provide additional details about the political history and objectives of the TRI, as well as how the program contributes to the enforcement of other laws. In sociology contexts, instructors may want to delve deeper into the policy's philosophical foundations in environmental equity and uses in grassroots activism

Instructor Preparation and Materials

Preparation for this exercise requires about 30 min. Two introductory presentations (slides in ESM-D and an online link to a Prezi presentation by the author provided in ESM-E) should be reviewed as possible resources to present to students. They resources outline the activity for creating social-ecological systems maps based on NPR's *Poisoned Places* series and include introductory talking points regarding the TRI and pollution management. The PowerPoint includes an introduction to social-ecological systems mapping, basic details about the TRI, and the NPR/systems mapping activity. The Prezi provides more comprehensive details about the political history and philosophical underpinnings of the TRI in addition to the social-ecological systems mapping activity. The online Prezi cannot be edited, though instructors can choose select parts to view in class. Instructors should use either the PowerPoint or Prezi, but not both.

Instructors should be aware of the basic elements of the TRI and social-ecological systems mapping prior to carrying out this activity. This understanding can be gained by reviewing the EPA's TRI Fact Sheets (ESM-A, ESM-B and ESM-C), the TRI website, and the example presentations (ESM-D and ESM-E). Additional background materials regarding the TRI and its philosophical foundations, social-ecological systems mapping, and industrial accidents are included in ESM-F. Two copies of the listening guide worksheet (ESM-G) should be printed for each student. Students will use this worksheet in class. The example presentations (ESM-D and ESM-E) include embedded links to two of the *Poisoned Places* case studies. Instructors should listen to/watch these prior to class and complete a listening guide for each. After reviewing these cases and others in the series, you may choose to use alternative stories in class, based upon unique objectives and students' interests.

Activities

1. Instructors should first provide an overview of the Toxics Release Inventory (12 min). Main talking points include:

    (a) The role of historic industrial accidents in raising the public's awareness about the risks of living near polluting facilities. Examples are provided in PowerPoint slide two.

    (b) The underlying purposes of community-right-to-know policies, which are to raise the public's awareness of risk and change, provide information that stakeholders can use to inform decision making, and enhance stakeholder participation in community planning (EPA 2015a). PowerPoint slide three highlights these purposes as they relate to EPCRA.

    (c) The TRI reporting criteria. Facilities must report to the TRI if they (EPA 2015a) (1) use at least one of the approximately 680 chemicals listed by EPA that pose chronic or acute health risks and/or carry risks to the environment, (2) meet designated industry classifications, and (3) have ten or more employees. These points are provided in PowerPoint slide four.

(d) Additional talking points and student-centered discussion questions are provided in ESM-H, and the supplementary presentations can be used as a road map for this introduction (see ESM-D and ESM-E).

2. Instructors should introduce social-ecological systems mapping as a tool for comprehensively conceptualizing the dynamic and interconnected natural and social elements of environmental conflicts and identifying leverage points for improving environmental governance among multiple stakeholders. You should also emphasize that while this activity focuses on community-scale conflicts related to industrial pollution, social-ecological systems mapping can be used to reveal patterns of behavior in many contexts. PowerPoint slide five emphasizes these points and outlines a four-step process for drawing social-ecological systems maps. The four-step process is intended to help students draw maps that contain four elements: nodes or parts, relationships, loops and thresholds, and leverage points (10 min).

(a) Providing students with an example map of a system with which they are already familiar can help build their conceptual understanding of system principles. The example I like to use, and that is included in PowerPoint slide six, is about drinking caffeine. ESM-D walks through a process for building a system map that illustrates behaviors related to caffeine intake in a fun and light-hearted way.

3. The purpose of the final portion of this activity is for students to practice social-ecological systems mapping by applying the four-step methodology to illustrate the dynamics of the TRI and other cases of hazardous industrial pollution at the community scale. Instructors should introduce *Poisoned Places* series, which documents case studies of stakeholders working at grassroots levels to reduce the risks from hazardous pollutants in their communities. This portion of the activity uses the NPR series as source material for students' social-ecological systems maps (25 min).

(a) Play the audio for the case study "*Secret Watch List Reveals Failure to Curb Toxic Air*" (Shogren 2011 and embedded in slide 7 of ESM-D). Students should fill out the listening guide (ESM-G) while listening to the story. Once the case study is over, instructors should lead the class through the creation of a social-ecological systems map on the blackboard (see an example on slide 8 in ESM-D which can be shown and used for comparison after the class creates its own). This process should be interactive, with students collaboratively constructing the map by describing the stakeholders, relationships, actions, and consequences they observed in the story as the instructor or a volunteer student draws the map on the board. Following this process, you can lead a class discussion about the process of drawing the maps. You may want to ask students which stakeholder relationships are more or less visible in the report and to identify leverage points that offer opportunities for managing the system to reduce conflict and pollution.

(b) The class should then watch a second *Poisoned Places* story, *Despite Warnings From Inspector, One Iowa Town Still Battles Toxic Air* (Berkes 2011, and included in slide 9 of ESM-D) and use the listening guide (ESM-G) to take notes. Following the video, students can work individually or in small groups to create social-ecological systems maps that illustrate the relationships noted in the documentary. An example social-ecological systems map of this case study is provided in slide 10 of ESM-D (again, the example should not be revealed until after students create their maps). The instructor should conclude with a discussion about the similarities and differences among students' system diagrams, emphasizing the inclusion of the four elements of systems maps. As a final synthesis activity, instructors should ask students to identify leverage points for managing the system and de-escalating the conflicts reported in the story. During this discussion, you may want to ask students to distinguish among leverage points used by stakeholders in the story and alternative leverage points that the students, from their perspectives as outsiders to the story, believe offer new opportunities for reducing conflict in this scenario.

Follow-Up Engagement

- As a follow-up assignment, instructors can ask students to listen to additional case studies from the NPR (2015) website and create social-ecological systems maps (that are described in more detail with narrative) that illustrate the governance relationships and possible leverage points embedded within them. A description of this assignment is provided in ESM-I.
- Instructors that wish to explore the politics of pollution could facilitate a class discussion about the durability of the TRI over time. Periodically the TRI comes under attack by members of the US Congress that support the reform of policies seen as limiting or slowing economic growth. Instructors can ask students to predict some of the arguments these policy makers might use to advance their agendas. Contrary to our assumptions, some of the TRI's biggest supporters are industry stakeholders. Instructors can ask students to think of some of the reasons industry groups might support the TRI and want to maintain its strength.
- EPA's TRI office has initiated a University Challenge (EPA 2015b) program to generate new ways of using, communicating, and visualizing TRI data. The goals of this program are to engage students in improving awareness and access to the online TRI databases and mapping tools. As a follow-up assignment, instructors may have students write proposals for this program. Details are found on the website (EPA 2015b).
- For a global perspective on toxic pollution, students can identify and explore community-right-to-know policies in other countries and compare them to the TRI.
- Activities to explore environmental justice are included in Chaps. 27–30 of this volume.

Connections

- The TRI website (http://www2.epa.gov/toxics-release-inventory-tri-program) provides unique mapping tools that are searchable by location and allow students to explore the distribution of polluting facilities across communities. Students can use these tools to explore their hometowns or other communities. These tools are connected to the US Census and enable connections to environmental justice and equity (e.g., Chaps. 27–30, this volume).
- Exploring EPA's risk screening tools, found at http://www.epa.gov/risk/, can provide insight and analysis into the particular risks that accompany specific chemicals as well as risks that are faced by sensitive populations such as children, pregnant women and those of child-bearing age, and the elderly. This allows the activity to be integrated into discussion of public health and epidemiology.
- Students interested in environmental education and outreach may be curious about the communication tools created by EPA and the effectiveness of these programs.

Acknowledgements The original research for this project was conducted in collaboration with Caitlin Briere and Nora Lopez in the Environmental Protection Agency's Toxics Release Inventory program. The project was supported by EPA's TRI University Challenge.

References

Berkes H (2011) Despite warnings from inspector, One Iowa Town still battles toxic air. National Public Radio. http://www.npr.org/blogs/thetwo-way/2011/11/30/142948573/despite-warnings-from-inspector-one-iowa-town-still-battles-toxic-air. Accessed 8 May 2015

Environmental Protection Agency (EPA) (2015a) Learn about the toxics release inventory. http://www2.epa.gov/toxics-release-inventory-tri-program/learn-about-toxics-release-inventory. Accessed 12 Jan 2015

Environmental Protection Agency (EPA) (2015b) 2015 TRI University challenge. http://www2.epa.gov/toxics-release-inventory-tri-program/2015-tri-university-challenge-0. Accessed 8 May 2015

National Public Radio (NPR) (2011–2013) Poisoned Places: toxic air neglected communities. http://www.npr.org/series/142000896/poisoned-places-toxic-air-neglected-communities. Accessed 12 Jan 2015

Shogren E (2011) Secret watch list reveals failure to curb toxic air. National Public Radio. Poisoned Places: Toxic Air, Neglected Communities. http://www.npr.org/2011/11/07/142035420/secret-watch-list-reveals-failure-to-curb-toxic-air. Accessed 8 May 2015

Toxics Release Inventory (TRI) (1986) 42 C.F.R. Chapter 116. §313. Emergency Planning and Community Right to Know Act. U.S. Code. http://www.gpo.gov/fdsys/pkg/USCODE-2011-title42/html/USCODE-2011-title42-chap116.htm. Accessed 29 July 2015

Chapter 32
The Skies, the Limits: Assessing the Benefits and Drawbacks of Tighter US Soot Emission Standards

Sandra L. Cooke

Introduction

Exposure to air pollution is linked to numerous health problems including asthma, cancer, low birth weight, and cognitive impairment (Grahame et al. 2014; Kim et al. 2013). The US Environmental Protection Agency (EPA) sets National Ambient Air Quality Standards (NAAQS) for six "criteria" pollutants that are particularly damaging to human health (EPA 2014), and the US Clean Air Act mandates that the EPA periodically review the NAAQS and the science upon which they are based (EPA 2013b). In December 2012, the EPA announced that it would tighten the limit on one criteria pollutant: fine particulate matter (PM 2.5). Commonly called soot, PM 2.5 is emitted by diesel engines, coal burning, and industrial activities (among other sources) and is associated with detrimental cardiovascular and respiratory conditions (Kelly and Fussell 2012). The tighter soot limits were praised by public health advocates but denounced by stakeholders in manufacturing and related industries (Shogren 2012). The National Association of Manufacturers (NAM) filed a legal challenge to the new standard, but in May 2014, a federal appeals court upheld the EPA's decision (Barboza 2014). The regulation took full effect in 2015.

The goal of the discussion-based activity described below is for students to think critically about this environmental health issue and assess the benefits and drawbacks of the new EPA limit. Additionally, after reading about current research on the chemical composition of particulate matter, students consider how an improved understanding of the specific components of PM 2.5 and their associated health effects could lead to more effective air quality regulations.

Electronic supplementary materials: The online version of this chapter (doi:10.1007/978-3-319-28543-6_32) contains supplementary material, which is available to authorized users.

S.L. Cooke (✉)
Department of Biology, High Point University, High Point, NC, USA
e-mail: scooke@highpoint.edu

© Springer International Publishing Switzerland 2016
L.B. Byrne (ed.), *Learner-Centered Teaching Activities for Environmental and Sustainability Studies*, DOI 10.1007/978-3-319-28543-6_32

Learning Outcomes

After completing this activity, students should be able to:

* Identify why soot, also known as fine particulate matter (or PM 2.5), is an EPA criteria pollutant that can be harmful to human health.
* Discuss two competing perspectives on the 2012 EPA soot regulation and argue in favor or opposition of the new policy.
* Explain how new scientific knowledge on the composition of soot could help develop better solutions for its regulation.

Course Context

* Developed for an introductory environmental science course with 20–50 students, mostly non-majors
* 50 min in one class meeting
* Students should have some background on NAAQS, as detailed below
* Could be adapted to smaller or larger classes and may also be appropriate for environmental policy or ethics courses

Instructor Preparation and Materials

Instructors should be familiar with the EPA NAAQS and the sources and impacts of PM 2.5 (soot). The EPA's websites (EPA 2013a, b; 2014) and review articles by Kelly and Fussell (2012) and Cassee et al. (2013) can be consulted for basic information needed to implement this lesson. A key article is that of O'Connor et al. (2008), who found that short-term increases in PM 2.5 to levels below NAAQS were associated with adverse respiratory health effects in inner city children in the USA. Instructors interested in additional details could peruse the EPA's Integrated Scientific Assessment (ISA) for Particulate Matter (US EPA 2009), which describes more of the science behind the EPA's decision to tighten soot standards (it is, however, a bit overwhelming, with over 1000 pages and thousands of references).

This activity is most effective if students have been introduced to some basic principles of air pollution (e.g., EPA criteria pollutants and their sources and effects; overview of the Clean Air Act and NAAQS). If such topics were not covered in the course before using this activity, instructors should consider providing students with additional introductory information (e.g., via a mini-lecture at the beginning of class or a pre-class reading or video) based on the context of their course. References cited above, along with general environmental science textbooks, can be consulted to develop a presentation that matches the specific context of a course. In addition, the EPA air quality website is a good resource for updates on regulations and visual aids (US EPA 2014).

Materials used in the activity should be reviewed prior to the class period, including the audio news story (Shogren 2012), three brief articles (ALA 2012; NAM 2012; Peake 2012; text of the latter provided in Electronic Supplementary Material (ESM)-A), a handout with discussion questions (ESM-B), presentation slides (ESM-C), and the instructor's guide for discussion (ESM-D). In the presentation slides, the instructor may wish to replace the slides containing information about PM 2.5 in Central North Carolina with that for their own region as can often be found on the website of a state's (or country's) division or department of air quality (or using other references provided below). Printed or pdf copies of the audio transcript and three articles should be provided to each student before class or for use in class. Assemble the five handouts in a single packet ordered as follows: discussion questions, audio transcript (Shogren 2012), ALA press release (ALA 2012), NAM press release (NAM 2012), and NC State research blog article (Peake 2012). To play the audio of the NPR story and have students read the online articles in class, an audio system and Internet access are required in the classroom, including for all students.

This activity involves discussion in groups of two to four students, so the instructor should decide on a method of group formation such as (1) using a deck of playing cards to randomly assign groups; (2) strategic assignment based on the instructor's knowledge of each student's personality, motivation, leadership qualities, etc.; (3) allowing students to choose their own groups; or (4) where students are sitting. The instructor should also decide if each student will turn in a completed handout individually (preferable to ensure engagement by all) or if each group will submit one.

Activities

This activity uses small-group and whole-class discussions in an "interrupted" format. First, the instructor briefly (3–4 min) reviews the six EPA criteria pollutants (PM, nitrous oxides, sulfur oxides, carbon monoxide, ozone, lead) and explains that NAAQS are updated periodically when environmental health research suggests that existing standards are inadequate (US EPA 2009, 2013b and O'Connor et al. 2008). Groups should then be formed based on the instructor's preferred method. After distributing the five handouts (ESM A and ESM B) and allowing the students some time to review Part 1 of the discussion guide, the instructor plays the NPR news story reporting on the December 2012 decision to tighten soot standards (Shogren 2012; listening to the story together can enhance students' grasp of the issue, although the transcript should still be provided to the students). After this, students are given 20–25 min to read two brief press releases (ALA 2012; NAM 2012) and answer the questions in Parts 1 and 2 of the handout in their small groups. The instructor should circulate through the room to answer questions, ensure that students stay on task, and encourage further discussion if necessary (e.g., for question 7 the instructor can ask individual students or the small group how strongly they support or do not support the new standard and what evidence or claims they considered in their decision—see ESM-D).

Before students begin Part 3 of the handout, the instructor should interrupt the group discussions to review with the class what PM 2.5 is using slide 3 (ESM-C), which shows how "fine" 2.5 µm actually is. Next, the instructor can poll the small groups to determine who supports and opposes the tighter standard, and whether each group's decision was unanimous. As desired and as time permits, this would be a good moment to engage the class with discussion questions about personal perceptions of the issue (Box 32.1, questions 1 and 2; ESM-D). Showing slide 4 (ESM-C) will inform students of national PM 2.5 pollution levels in the USA from 2000 to 2013 which can be paired with question 3 in Box 32.1. The instructor should then explain slide 5, which is from the research blog post that students will read for Part 3 (ESM-C). It shows the relationship between elemental carbon—a component of PM 2.5—and cardiovascular disease hospitalization risk. Then, students should be given about 10–15 min to read the blog post (Peake 2012) and discuss answers to Part 3 of the handout as the instructor moves around the room again.

With approximately 5–10 min of class remaining, the instructor brings the whole class together to discuss question 10 of the student handout (how more research could lead to improved air quality regulation). Initially, students may answer this question by saying research will show that PM 2.5 is not as bad as we thought and therefore does not need stringent regulation. However, if they carefully read, think about, and discuss the research blog post, they should see the connection between understanding which specific components of PM 2.5 are most harmful and developing more nuanced NAAQS that target only those harmful components (ESM-D). If time allows, the instructor can conclude the activity by briefly explaining that some regions have a history of non-attainment for the previous PM 2.5 standard (in ESM-C, slides 6–8 show NC as an example), which implies that there could be problems meeting the new, tighter standard. The instructor could then ask discussion of question 4 (Box 32.1; ESM-D).

Box 32.1. Questions to Enhance Discussion

1. Do you think our class poll results would mirror that of the American public? That is, if we were to poll the US (or North Carolina) public on their opinion of the new soot standard, what would the results look like?
2. How do individuals' values, worldviews, and experiences influence their perception of this issue and of environmental issues more broadly?
3. Does the recent trend in PM 2.5 affect how strongly you do or do not support the new standard?
4. The new soot limit has been in place since December 2012, but mitigation measures begin in 2015 and should be fully implemented by 2020. What type of monitoring or research should be done to determine if the tighter limit is more effective than the previous limit?

Follow-Up Engagement

- The following are possible prompts for assignments or assessment:
  - Select one EPA criteria pollutant, identify one of its main sources, and describe its effects on the environment and/or human health.
  - Explain the benefits and drawbacks of the new EPA standard on fine particulate matter.
  - Write a letter to the editor or op-ed article about the EPA's decision to tighten the standard and the more recent ruling by a federal appeals court to uphold the new regulation (e.g., Barboza 2014; activity described in Chap. 40, this volume).
  - Write a press release of the study by O'Connor et al. (2008) that documented detrimental respiratory health effects in children exposed to PM 2.5 levels below the former NAAQS.
  - Design a poster for a public museum exhibit that explains how soot is hazardous to human health.
  - Investigate how other countries have dealt with PM 2.5 regulations (or with air pollution more generally). Resources from the World Health Organization (WHO) may be a useful starting point (e.g., WHO 2014).

Connections

- This activity addresses how trade-offs are often made when crafting public policy (e.g., Chap. 4, this volume). False dichotomies of "humans vs. the environment" or "the environment vs. the economy" are often used to frame environmental problems. While claims are made that tighter regulations will stifle job growth, one could also point to the economic costs of detrimental public health effects. Arguably, this activity exemplifies that more sustainable solutions (those with a long-term, more holistic outlook) ultimately benefit the environment, the economy, and, thus, society as a whole (i.e., the triple bottom line; see Chaps. 23 and 24, this volume).
- Regulating air pollution is often considered an environmental justice issue (see Chaps. 27–30, this volume). O'Connor et al. (2008) and references therein provide detailed information on the increased vulnerability of poor urban residents to adverse effects of air pollution.
- The US Clean Air Act gives the EPA considerable authority in regulating criteria pollutants, which some may view as excessive. Hence, this activity is relevant to the broader issue concerning the government's role in environmental policy.

References

American Lung Association (ALA) (2012) American Lung Association applauds EPA decision to protect public from soot. http://www.lung.org/press-room/press-releases/new-soot-standard-2013.html. Accessed 29 Jul 2015

Barboza T (2014) Obama administration limits on soot pollution upheld by appeals court. http://www.latimes.com/science/sciencenow/la-sci-sn-soot-air-pollution-epa-ruling-20140509-story.html. Accessed 5 Mar 2015

Cassee F, Héroux M, Gerlofs-Nijland M et al (2013) Particulate matter beyond mass: recent health evidence on the role of fractions, chemical constituents and sources of emission. Inhal Toxicol 25:802–812

Grahame T, Klemm R, Schlesinger R (2014) Public health and components of particulate matter: the changing assessment of black carbon. J Air Waste Manag Assoc 64:620–660

Kelly F, Fussell J (2012) Size, source and chemical composition as determinants of toxicity attributable to ambient particulate matter. Atmos Environ 60:504–526

Kim K, Jahan S, Kabir E (2013) A review on human health perspective of air pollution with respect to allergies and asthma. Environ Int 59:41–52

National Association of Manufacturers (NAM) (2012) EPA doubles down on regulations; manufacturers and the economy lose. http://www.nam.org/Communications/Articles/2012/12/EPA-Doubles-Down-on-Regulations-Manufacturers-and-the-Economy-Lose.aspx. Accessed 29 Jul 2015

O'Connor G, Neas L, Vaughn B et al (2008) Acute respiratory health effects of air pollution on children with asthma in US inner cities. J Allergy Clin Immunol 121:1133–1139

Peake T (2012) In particulate matter, the particulars matter. https://news.ncsu.edu/2012/10/tp-particulatematter/. Accessed 29 Jul 2015

Shogren E (2012) EPA targets deadliest pollution: soot. http://www.npr.org/2012/12/17/167427988/epa-targets-deadly-soot-pollution. Accessed 29 Jul 2015

U.S. Environmental Protection Agency (EPA) (2009) Integrated science assessment for particulate matter (Final Report). U.S. Environmental Protection Agency, Washington, DC, EPA/600/R-08/139F

U.S. Environmental Protection Agency (EPA) (2013a) Particulate matter. http://www.epa.gov/airquality/particlepollution/. Accessed 4 Nov 2014

U.S. Environmental Protection Agency (EPA) (2013b) Process of reviewing the National Ambient Air Quality Standards. http://www.epa.gov/ttn/naaqs/review.html. Accessed 9 Mar 2015

U.S. Environmental Protection Agency (EPA) (2014) Six common air pollutants. http://www.epa.gov/airquality/urbanair/. Accessed 25 Feb 2015

World Health Organization (WHO) (2014) Air quality deteriorating in many of the world's cities. http://www.who.int/mediacentre/news/releases/2014/air-quality/en/. Accessed 9 Mar 2015

Chapter 33
Don't Blame the Trees: Using Data to Examine How Trees Contribute to Air Pollution

Patrick W. Crumrine

Introduction

In 1980, US President Ronald Reagan stated that "approximately 80 % of our air pollution stems from hydrocarbons released by vegetation, so let's not go overboard in setting and enforcing tough emission standards from man-made sources" (Pope 1980). This statement was widely criticized by environmentalists because it oversimplified and exaggerated the contribution of plant emissions to air pollution. While plants do not cause pollution directly, they do emit approximately 75–80 % of the volatile organic compounds (VOCs) in the USA (Lerdau and Slobodkin 2002; US EPA 2013). The remaining 20–25 % comes from various industrial processes and vehicular emissions. In the presence of sunlight, VOCs react with nitrogen oxides (hereafter NO_x), produced from fuel combustion, to form ground-level ozone (Purves et al. 2004). Ground-level ozone is a greenhouse gas, a human health hazard, and negatively affects plant productivity resulting in billions of dollars in crop losses globally (Ainsworth et al. 2012). Since the industrial revolution, baseline ground-level ozone concentration has increased from ~10 ppb to 35–40 ppb in many parts of the Northern Hemisphere, and ozone pollution events during the summer, with levels exceeding 70 ppb (the current US EPA standard), are a routine occurrence in both

Electronic supplementary materials: The online version of this chapter (doi:10.1007/978-3-319-28543-6_33) contains supplementary material, which is available to authorized users.

P.W. Crumrine (✉)
Department of Geography and Environment, Rowan University, Glassboro, NJ, USA

Department of Biological Sciences, Rowan University, Glassboro, NJ, USA
e-mail: Crumrine@rowan.edu

© Springer International Publishing Switzerland 2016
L.B. Byrne (ed.), *Learner-Centered Teaching Activities for Environmental and Sustainability Studies*, DOI 10.1007/978-3-319-28543-6_33

urban and rural regions (Hartmann et al. 2013). Although anthropogenic emissions of some ozone precursors such as NO_x have likely declined in the last 20–30 years in Europe and North America (Hartmann et al. 2013), significant amounts of NO_x and other precursor chemicals are still emitted and lead to the formation of ground-level ozone. Considering the contribution of plants to this problem is important because plant VOCs are highly reactive and are so abundant in some regions that reductions in anthropogenic VOCs have had little impact on ground-level ozone concentrations. Quantifying the relative contribution of plant VOCs to the production of ground-level ozone can inform approaches to minimizing air pollution.

In this chapter's activity, students will become familiar with factors influencing the production of ground-level ozone with an emphasis on the role of VOC-emitting trees in North America. After completing a background reading about air pollution concepts, students interpret data that document the increase in ground-level ozone between 1950 and 2010, the relative contribution of natural and anthropogenic ozone precursors, and the spatial extent of ground-level ozone pollution in the USA. In groups, students answer questions about the data and synthesize the information to conclude the activity. The larger goal of this activity is to help students develop skills in interpreting and synthesizing data to explain the formation of an important secondary air pollutant (ground-level ozone) and recognize that effective environmental policy requires sound scientific understanding of environmental processes.

Learning Outcomes

After completing this activity, students should be able to:

- Distinguish between primary and secondary air pollutants and identify precursor chemicals that lead to the formation of ground-level ozone.
- Interpret figures and tables that document temporal and spatial changes in the concentration of ground-level ozone and its precursors.
- Explain the relative contributions of anthropogenic and natural emissions to the formation of ground-level ozone.
- Evaluate different management approaches for reducing air pollution.

Course Context

- The in-class portion of the full activity requires 75–100 min and may extend across multiple class sessions
- Students acquire background knowledge on air pollution through assigned readings

- This activity was developed for an introductory environmental science course for environmental studies majors and is implemented toward the end of the unit on air pollution. This activity works best in relatively small classes (<36), with students working in groups of three to five
- A slightly abbreviated version of the activity that omits Table 2 and Fig. 4 requires 55–70 min. It can also be adapted for larger classes with the instructor leading all students through the activity or as a homework assignment

Instructor Preparation and Materials

Prior to implementing the activity, the instructor should become familiar with the key concepts underlying this activity (ESM-A). Reading the papers and reports that the figures and tables for the activity were obtained from will provide additional context and background; citations are included with each figure in ESM-B. Instructors should also review the answers to questions assigned to students (ESM-C). In the class session prior to the activity, the instructor should assign the chapter on air pollution if a textbook is used in the course. Alternatively, *The Plain English Guide to the Clean Air Act* (US EPA 2007) provides a good introduction to key concepts. These readings should familiarize students with the basic concepts associated with air pollution pertinent to the activity such as anthropogenic sources of air pollutants, their environmental and health effects, and the basic principles of the Clean Air Act. Students should also recognize the difference between "good" ozone in the stratosphere (i.e., the ozone layer) and "bad" ground-level ozone. Instructors may choose to prepare a lecture or activity to review key terms and concepts (ESM-A) with students before the activity. Alternatively, a quiz (e.g., ESM-D) at the start of the activity may provide additional incentive for students to complete the assigned reading. The instructor should review, edit as desired, and print enough copies of the figures and tables (ESM-B) and the worksheet (ESM-E) to distribute to each student at the beginning of the activity. Alternatively these could be made available through an online course management platform for students to access and complete electronically. The instructor should be prepared to divide students into groups of three to five depending on the size of the class.

If instructors have less time for the activity, Table 2 and Fig. 4 (in ESM-B) can be removed. If this modification is made, it is important for instructors to help students recognize that a majority of the non-methane VOCs emitted are from natural sources, mostly isoprene from trees. This information can be gleaned from Table 1. The fact that isoprene emission is species specific (Table 2) is interesting but not a necessary detail to understand the basic concept. Similarly, the change in isoprene emission in the Eastern USA (Fig. 4) is not absolutely necessary for working with the other data.

Activities

Instructors can begin the activity with the famous Reagan quote stated in the introduction (or "Trees cause more pollution than automobiles do" which has also been attributed to Reagan). Initiate a short discussion (e.g., 5–10 min) by asking students whether or not there is any truth to these statements. In the author's experience, most students respond with a resounding "no," which is not surprising considering that many introductory textbooks do not mention the emission of VOCs by plants. To follow up, ask students how plants influence the composition of the atmosphere (e.g., by reducing CO_2 and increasing O_2 levels). It may also be helpful to ask students if plants have any distinctive odors or fragrances because these scents are also caused by VOCs. The key concept to illuminate is that plants exchange gasses with the atmosphere. After this discussion (and other introductions or a quiz as desired), the instructor then asks students to form groups for the activity.

To facilitate the activity—and depending on the level and background of the students—instructors might find it helpful to start by explaining the pattern in the first figure to the entire class, pointing out things like the axis titles, units, and scale. Students should then spend 5–10 min considering the information in each figure/table (ESM-B) and answering the worksheet questions (ESM-E). The figures and tables are sequenced so that students (1) document the existence of ground-level ozone pollution, (2) identify temporal and spatial patterns in the emission of ozone precursor chemicals, (3) identify tree species with high isoprene emission potential, (4) identify changes in plant cover that contribute to isoprene emission, and (5) synthesize multiple types of data to explain ground-level ozone pollution events. On the worksheet, the first question for each figure or table asks students to explain the pattern or relationship. Subsequent ones ask them to suggest mechanisms for the pattern and consider the environmental consequences. For all questions, each student group should write a brief response on the worksheet that can be submitted to the instructor as desired.

As students complete the worksheet, the instructor should walk around the room to gauge progress, answer questions, and help students if they are having difficulty. Instructors may choose to implement "checkpoints," where students stop their small group discussions and the instructor concisely describes the patterns that students should have identified from the data. In particular, a checkpoint after Table 1 can be used to reinforce the fact that ground-level ozone and its precursors have increased over time and there are both natural and anthropogenic sources. For Fig. 4, a key discussion point to share is that some trees emit significant quantities of isoprene (a VOC), which are broadly distributed in the USA and have increased in recent years. In the author's experiences, students are usually able to glean basic information from the data about levels of pollutants in the atmosphere and emissions from natural and anthropogenic sources. However, students tend to have more difficulty synthesizing the information to predict where ground-level ozone pollution should occur and understanding that pollution control must take into account both natural and anthropogenic sources. To help students understand these concepts and bring the activity to a close, it is helpful to discuss answers to the concluding question in

ESM-E with the entire class: Considering what you now know about isoprene emitted from trees, how do you interpret the statements made by Ronald Reagan in the early 1980s? Potential answers to this and the other questions and additional commentary that can be conveyed to students are located in ESM-C.

Follow-Up Engagement

- These prompts could be used for extended discussion or assessment questions (potential responses and additional commentary are located in ESM-F):

  - Summarize the role of anthropogenic and biogenic emissions in the formation of tropospheric ozone.
  - In urban regions, ozone production tends to be limited by the availability of VOCs, while in rural regions, it is limited by the availability of NO_x. Why do you think this is the case?
  - Suppose you were responsible for coordinating a large-scale tree-planting program in an urban region such as New York City. What species would you select to plant and why?

- Have students explore the impact of the nonnative invasive chestnut blight on forest composition in North America, how this may have influenced isoprene emission from these forests, and ultimately its effects on ground-level ozone (Lerdau et al. 1997).
- A similar exercise would be for students to explore how different forest management practices influence isoprene emission and ground-level ozone production. For example, planting high isoprene-emitting tree species for biofuel production has the potential to influence ground-level ozone production (Ashworth et al. 2013).
- Encourage students to visit the AirNow (http://airnow.gov/) website maintained by the US EPA to see the air quality index (AQI) for their home region.

Connections

- Ozone is one of the more misunderstood components of our atmosphere because it has positive effects in the stratosphere by screening out harmful UV radiation but several negative environmental effects at the ground level (troposphere). This activity helps to reinforce basic concepts related to atmospheric composition and structure and illuminates how human activity interacts with natural processes to change that composition, influencing global climate and human health.
- Controlling ground-level ozone pollution is a relatively complex task that requires an understanding of both natural and anthropogenic emissions and consideration of politics, economics, technology, and other social variables. Thus, this activity can be referred to when discussing many topics in an interdisciplinary class that explores relationships among science, the environment, and society.

References

Ainsworth EA, Yendrek CR, Sitch S, Collins WJ, Emberson LD (2012) The effects of tropospheric ozone on net primary productivity and implications for climate change. Annu Rev Plant Biol 63:637–661

Ashworth K, Wild O, Hewitt CN (2013) Impacts of biofuel cultivation on mortality and crop yields. Nat Clim Change 3:492–496

Hartmann DL, Klein AMG, Tank M et al (2013) Observations: atmosphere and surface. In: Stocker TF, Qin D, Plattner GK et al (eds) Climate change 2013: the physical science basis. Contribution of working group I to the fifth assessment report of the Intergovernmental Panel on Climate Change. Cambridge University Press, Cambridge, pp 159–254

Lerdau M, Slobodkin L (2002) Trace gas emissions and species-dependent ecosystem services. Trends Ecol Evol 17:309–312

Lerdau M, Guenther A, Monson R (1997) Plant production and emission of volatile organic compounds. Bioscience 47:373–383

Pope C (1980) The candidates and the issues. Sierra 65:15–17

Purves DW, Caspersen JP, Moorcroft PR, Hurtt GC, Pacala SW (2004) Human-induced changes in US biogenic volatile organic compound emissions: evidence from long-term forest inventory data. Global Change Biol 10:1737–1755

US EPA (2007) The plain english guide to the clean air act. http://www.epa.gov/airquality/peg_caa/pdfs/peg.pdf. Accessed 27 May 2015

US EPA (2013) Overview of the air pollutant emissions in the 2011 national emissions inventory version 1.01. http://www.epa.gov/ttn/chief/net/lite_finalversion_ver10.pdf. Accessed 12 Jan 2015

Part V
Information Literacy and Communication

Part V
Information Literacy and Communication

Chapter 34
Evaluating the Effectiveness of Green Roofs: A Case Study for Literature Research and Critical Thinking

Erika Crispo

Introduction

Green roofs are installed on buildings to ameliorate urban environmental problems including the heat island effect, water runoff, air pollution, and biodiversity loss (Oberndorfer et al. 2007; Rosenberg 2012). While the popularity of green roofs has increased in recent years, questions have been raised about their efficacy for addressing these issues. For example, Kraft (2013) argued (in *Scientific American*) that green roofs are ineffective as currently designed; this contrasts with a more positive opinion published less than a year earlier in *The New York Times* (Rosenberg 2012). This contrast begs the questions: Why is there a discrepancy in viewpoints? Which article should we believe? Insights into contentious issues, such as the efficacy of green roofs, can be gained through exercises that evaluate evidence from published research (also see Chaps. 9 and 15, this volume).

The goal of the activity described below is to help students develop science literacy skills by comparing claims made in the popular media with information from primary research articles. To evaluate the popular articles cited above, students search the primary scientific literature for evidence to support or refute the articles' different views about the environmental costs and benefits of green roofs. This exercise provides an opportunity for students to improve their analytic and critical thinking skills. In addition, analyzing green roofs integrates diverse environmental

Electronic supplementary materials: The online version of this chapter (doi:10.1007/978-3-319-28543-6_34) contains supplementary material, which is available to authorized users.

E. Crispo (✉)
Department of Biology, Pace University, New York, NY, USA
e-mail: ecrispo@pace.edu

© Springer International Publishing Switzerland 2016
L.B. Byrne (ed.), *Learner-Centered Teaching Activities for Environmental and Sustainability Studies*, DOI 10.1007/978-3-319-28543-6_34

concepts in context of a single management solution, fostering skills for synthesizing information. This case study framework also allows students to evaluate relationships between environmental problems and proposed solutions.

Learning Outcomes

After completing this activity, students should be able to:

- Identify urban environmental problems and their causes.
- Search for and analyze scientific journal articles and evaluate the quality of research.
- Critically evaluate claims made in popular media articles using information obtained from the primary literature.
- Discuss the pros and cons of a potential environmental management solution.
- *Optional advanced outcome*: Design experiments to test hypotheses.

Course Context

- Developed for an upper-year environmental science course for science majors, with 8–20 students
- 3 h in one class meeting
- Before class, students read two news articles and answer homework questions
- Suitable for any environmental science or studies course; for shorter class sessions, the literature search activity can be performed as homework prior to the class meeting (see Box 34.1)

Box 34.1

If students do not have access to wireless internet or laptops/tablets during class time, the activity can be modified. Instead of assigning the news articles (Rosenberg 2012; Kraft 2013) and questions (ESM-B) as homework, they can be assigned in class with students answering the questions in small groups. The instructor walks around the room during the activity to ensure that students stay on track and to answer any questions the students may have. After an in-class discussion, the students are assigned the literature search for homework. (For out-of-class work, it may be best to assign work to be done individually rather than in groups, to avoid the problem of one student performing most of the work.) Assign each student one of the four problems that

(continued)

> **Box 34.1.** (continued)
>
> green roofs are intended to alleviate (see main text), with the number of students assigned each problem allocated equally. Students are asked to select two relevant primary research articles on their topic, summarize the key findings, analyze the quality of the research study (e.g., did they use an adequate representation of control and experimental groups?), and evaluate the effectiveness of green roofs based on their findings. Students could be asked to write a report about their findings, or present results in the next class meeting, with detailed instructions developed by the instructor.

Instructor Preparation and Materials

Begin by reading the two news articles that students are assigned for homework (Rosenberg 2012; Kraft 2013). Instructors should search for and read the literature to see which articles students might retrieve and to gain background information about green roofs. Background reading should also include studies referred to in the two news articles (Oberndorfer et al. 2007; Gaffin et al. 2010, 2011; MacIvor and Lundholm 2011; McGuire et al. 2013). Additional articles are listed in Electronic Supplementary Materials (ESM)-A. Key points to consider are the effectiveness of green roofs for reducing the heat differential between the inside of buildings and the outside ambient temperature, how much water green roofs can absorb, whether green roofs can harbor native biodiversity, and how well green roofs remove pollutants from the air.

Before the class session, students read the two news articles (Rosenberg 2012; Kraft 2013) and answer the questions provided in ESM-B, with modifications, additional instructions, and context provided at the instructor's discretion. The questions are designed to ensure comprehension of the readings and to guide the students on what to focus on.

For the in-class activity, students should be told to bring their laptops or electronic tablets. Wireless internet will be needed in the classroom for the literature search; if it is not available, an alternative approach is outlined in Box 34.1, or instructors can distribute hard copies of relevant articles to the students.

Activities

Begin with a discussion of the homework questions (ESM-B) so students understand the content of the articles, the controversy surrounding green roofs, and the goal of the in-class activity. Convey to students that the purpose of the exercise is to acquire skills in attaining information to draw scientifically informed decisions on contentious issues. During class time, students form small groups of two to four students, with each group assigned one of four topics (as identified by the readings: the heat island effect, water runoff, air pollution, and biodiversity loss). If multiple groups are assigned the same topic, effort should be made to assign each topic to an equal number of groups. In groups, the students then perform an online literature search (e.g., using a library search engine or Google Scholar) to find peer-reviewed articles on their assigned topic (30–60 min). If possible, students should download the entire articles to be able to glean detailed information from results sections (including figures and tables). In some cases, the full article might not be readily available, so students must rely on information summarized in the abstract. (To facilitate students' more detailed evaluations, instructors could prepare hard copies of inaccessible full articles for students to consult after they locate abstracts.) As students conduct their searches, the instructor should walk around to make sure that students are on the right track and offer advice and encouragement.

Depending on the class size and amount of time available, the activity can proceed in one of two ways. For smaller classes and/or those with longer meeting times, student groups can present their findings to the whole class (e.g., using slides or a dry-erase board, 5–10 min per group). For larger and/or shorter class sessions, groups can summarize their findings in short, written reports after the class meeting (with instructions provided at the instructor's discretion). An example of results that can be expected from this activity is included in Box 34.2.

Box 34.2

As new research is published, general conclusions about the effectiveness of green roofs may change. In the fall of 2014, the author's class found support for green roofs on the grounds that they act as heat insulators in the winter and "air conditioners" in the summer (e.g., Oberndorfer et al. 2007; Getter et al. 2011) and reduce rainwater runoff (e.g., Mentens et al. 2006; Oberndorfer et al. 2007). Research was still inconclusive as to whether green roofs are capable of hosting thriving biotic communities (e.g., Williams et al. 2014). Students were able to identify problems associated with certain studies, such as the lack of replication of treatments and controls among several roofs, which could be problematic if green roofs were shaded and control roofs received sun, for example.

The most difficult point for students to evaluate was the role of green roof plants in taking up carbon dioxide pollution. A common misconception held by students was that plants take up carbon from the soil instead of from the air. When asked to design an experiment to examine this, students initially proposed using complex airtight greenhouses to test whether green roofs are sequestering carbon dioxide from the air. The instructor then clarified that the

(continued)

Box 34.2. (continued)

dry weight of plants is composed primarily of carbon and that *all* of this carbon comes from the air through the process of photosynthesis. Thus, whether plants on green roofs are sequestering carbon depends only on whether these plants are alive and growing. Of course, the carbon sequestration of the plants must offset the air pollution "footprint" associated with building the green roof structure for them to have a net positive effect on carbon sequestration. A recent study found that the air pollution costs and benefits of green roof materials can be balanced after 13–32 years of use but that other environmental costs associated with plastics in green roof materials are not counterbalanced (Bianchini and Hewage 2012).

To conclude the class, lead a follow-up discussion to draw conclusions about the discrepancies between the two news articles. Possible questions to guide this discussion include:

- Why do two articles, published only a year apart (Rosenberg 2012; Kraft 2013), come to such starkly contrasting conclusions?

  - Did new research published within that year affect Kraft's conclusion?
  - Was pivotal research missed by one of the authors?
  - Do the authors have inherent biases or different interpretations of the research?

- What do you conclude about the effectiveness of green roofs in ameliorating:

  - The heat island effect in the summer and heat loss in the winter.
  - Rainwater runoff.
  - Air pollution.
  - Habitat/biodiversity loss.

- Would the function of green roofs be expected to vary among cities with different climates (e.g., tropical vs. temperate)?

Follow-Up Engagement

- Students could choose one of the four "urban environmental problems" and design experiments to test the hypothesis that green roofs can be used to alleviate these problems. This could be done in groups in class or as a take-home assignment. Students should be asked to include (1) appropriate experimental and control groups, (2) an adequate level of replication, and (3) a clear explanation of how the outcome of the experiment will lead them to support or refute their hypothesis.

- A follow-up study could examine the costs of the materials to build a green roof and compare these costs to the expected benefits through savings from reduced heating and air conditioning.
- For activities related to urban sustainability, see Chaps. 17, 19, 20 and 30 of this volume.

Connections

- Green roofs relate to energy issues and carbon dioxide emissions because of how they can reduce energy use in buildings and sequester carbon. More broadly, these issues connect to discussions of climate change and management responses to it.
- Green roofs can be linked to topics on urban stormwater runoff (e.g., Chap. 20, this volume), eutrophication (e.g., Chap. 13, this volume), and soil erosion because of their roles in capturing rain water.
- The value of rooftop green spaces for pollinators (e.g., Tonietto et al. 2011) and native plants (e.g., MacIvor and Lundholm 2011) can be included in lessons on biodiversity and habitat conservation.

References

Bianchini F, Hewage K (2012) How "green" are the green roofs? Lifecycle analysis of green roof materials. Build Environ 48:57–65

Gaffin SR, Rosenzweig C, Eichenbaum-Pikser J et al (2010) A temperature and seasonal energy analysis of green, white, and black roofs. Columbia University Center for Climate Systems Research. http://www.coned.com/newsroom/pdf/Columbia%20study%20on%20Con%20Edisons%20roofs.pdf. Accessed 18 July 2015

Gaffin SR, Rosenzweig C, Khanbilvardi R et al (2011) Stormwater retention for a modular green roof using energy balance data. Columbia University Center for Climate Systems Research. http://www.coned.com/newsroom/pdf/Stormwater_Retention_Analysis.pdf. Accessed 18 July 2015

Getter KL, Rowe DB, Andresen JA, Wichman IS (2011) Seasonal heat flux properties of an extensive green roof in a Midwestern U.S. climate. Energ Build 43:3548–3557

Kraft A (2013) Why Manhattan's green roofs don't work—and how to fix them. Scientific American. http://www.scientificamerican.com/article/why-manhattans-green-roofs-dont--work-how-to-fix-them/. Accessed 18 July 2015

MacIvor JS, Lundholm J (2011) Performance evaluation of native plants suited to extensive green roof conditions in a maritime climate. Ecol Eng 37(3):407–417

McGuire KL, Payne SG, Palmer MI et al (2013) Digging the New York City skyline: soil fungal communities in green roofs and city parks. PLoS One 8, e58020

Mentens J, Raes D, Hermy M (2006) Green roofs as a tool for solving the rainwater runoff problem in the urbanized 21st century? Landscape Urban Plan 77:217–226

Oberndorfer E, Lundholm J, Bass B et al (2007) Green roofs as urban ecosystems: ecological structures, functions, and services. Bioscience 57:823–833

Rosenberg T (2012) Green roofs in big cities bring relief from above. The New York Times. http://opinionator.blogs.nytimes.com/2012/05/23/in-urban-jungles-green-roofs-bring-relief-from-above/. Accessed 18 July 2015

Tonietto R, Fant J, Ascher J et al (2011) A comparison of bee communities of Chicago green roofs, parks and prairies. Landscape Urban Plan 103:102–108

Williams NSG, Lundholm J, MacIvor JS (2014) Do green roofs help urban biodiversity conservation? J Appl Ecol 51:1643–1649

Chapter 35
The Story of Source Reliability: Practicing Research and Evaluation Skills Using "The Story of Stuff" Video

J.J. LaFantasie

Introduction

Reliance on simple Internet searches for quick access to information has increased dramatically in recent years. Because the Internet abounds with excellent sources, the general expectation that reliable information can be acquired easily has confounded development of research skills and the ability to determine source credibility (Metzger et al. 2003). This is a concern because, depending on the topic, much Internet-based information may come from agenda-driven groups and be of questionable authority (Stapleton 2005; Chap. 36, this volume). In particular, the polarizing nature of many topics within environmental science and sustainability studies has yielded a plethora of inaccurate, biased, and even falsified, web sources that may attract individuals who tend to utilize the most accessible source or the sources that agree most with their prior beliefs and biases (Metzger et al. 2010; White 2013). This may result in the use and propagation of misleading or inaccurate information. Tension between balanced and skewed representations of information has complicated research for students and the general public (Mandalios 2013; Stapleton 2005). Although students of the "Google generation" are assumed to be fluent in the digital world, their difficulty with determining bias and how to evaluate websites as sources of information are growing concerns for educators (McClure and Clink 2009; Metzger et al. 2003).

Electronic supplementary materials: The online version of this chapter (doi:10.1007/978-3-319-28543-6_35) contains supplementary material, which is available to authorized users.

J.J. LaFantasie (✉)
Colorado State University Agriculture Experiment Station, Grand Junction, CO, USA

Department of Biological Sciences, Fort Hays State University, Hays, KS, USA
e-mail: JJLaFantasie@fhsu.edu

L.B. Byrne (ed.), *Learner-Centered Teaching Activities for Environmental and Sustainability Studies*, DOI 10.1007/978-3-319-28543-6_35

"The Story of Stuff" (Leonard 2007), a popular web-based video focusing on relationships among economics, overconsumption, and environmental destruction, provides an exemplary case study for students to practice assessing source reliability and differentiating between informative and persuasive sources. It has received millions of views and has been used in numerous P-16 classes. Leonard's powerful message, delivery, and philosophy resonate with many and have has done much to raise awareness of the environmental impacts of waste and overconsumption. However, the presence of inconsistencies, bias, scare tactics, and misrepresentations undermine its usefulness as a fully reliable source of information. As such, this video serves as a focus for the activity described below which seeks to help students to develop their critical thinking skills and abilities to discern a source's credibility. Such skills are needed to help ensure that they use informed approaches to learn about and engage with complex environmental and sustainability issues.

Learning Outcomes

After completing this activity, students should be able to:

- Determine the reliability of a potential source of information.
- Distinguish between persuasive and informative sources.
- Express the importance of using credible sources to inform one's opinions.

Course Context

- Developed for a general education biological science course in environmental studies with 20–40 students
- Two 50 min class meetings
- Students conduct research to present in class, but no additional background is needed
- Adaptable to any class size and level that requires student critical thinking in research; it can be shortened to one class with an optional follow-up homework assignment

Instructor Preparation and Materials

The instructor should be familiar with the "Story of Stuff" video (Leonard 2007) and may find it useful to read or view critiques of the video (Doren 2009; Kaufman 2009) and critiques of critiques (Mark 2009). Additional references and suggestions

Table 1 Steps for critically evaluating information using the RADAR approach (modified from Mandalios 2013)

Relevance	Is the information relevant to your research?
Authority	Who is the author, and do they have the credentials necessary to provide information on the topic?
Date	How recently was the information posted?
Appearance	Does the information appear to be professionally produced?
Reason for writing	Did the author produce the piece to provide information, to advocate for an idea or product, or other purpose?

are provided in Electronic Supplementary Materials (ESM)-A. The instructor should also be familiar with, and provide to students, guidelines for how to critically evaluate the credibility and reliability of web-based sources (see Table 1, Mandalios (2013) and ESM-A for background and guidance). Although credibility and reliability can be distinguished, both are measures of "trustworthiness" addressed by the criteria presented in Table 1 and are used interchangeably for purposes of this activity. It is helpful if students have hard or electronic copies of Table 1 or other steps for evaluation of sources (see ESM-A or ESM-D) prior to commencement of the activity. A presentation prepared by the author contains many of the sources discussed above and in section "Activities" and may be useful to the instructor and students (ESM-D). In the classroom, a platform for web video presentation will be necessary to implement this activity without modifications.

This lesson is divided into two parts; for shorter adaptation, part 2 could be eliminated or used as an out-of-class writing assignment. In part 1 (50 min), students watch, critique, and discuss the "Story of Stuff." Alternatively, the video and evaluation could be assigned as a pre-class activity to allow more time for in-class discussion. ESM-D provides a work sheet that may facilitate part 1 activities. Part 2 includes an outside-of-class research assignment in which students investigate the validity of a claim (factoid) presented in the video and prepare for an in-class presentation in which they share their findings (requiring another class period; see ESM-D). For this research, the instructor may wish to assign factoids to be researched to avoid duplication among groups; examples are provided in ESM-B.

Activities

Part 1: Critique of Story of Stuff

At the beginning of class, the instructor could provide introductory talking points and questions that cause students to reflect on the importance of obtaining reliable information prior to decision making. A presentation prepared by the author (ESM-D) may be useful. For example, the dihydrogen monoxide example is compelling and speaks to the desire for avoidance of embarrassment associated with professing uninformed opinion (see ESM-B). In addition, students could be asked to reflect on a situation where uninformed or poorly informed decisions have negatively influenced their lives.

Following the introduction, the instructor should assign or have students assemble themselves into groups of two or three and provide each group with a list of steps for determining the reliability of web-based sources (Table 1, ESM-A, ESM-D). Groups should briefly examine the criteria (3–5 min) and then, as a class, review the steps to ensure understanding of what they are expected to assess (e.g., discuss the meanings of key terms in the table). This would also be a good time to briefly explain that determining the credibility of a source is important regardless of whether individuals agree with the premise or not, as this distinction can be difficult for some students. Additional details to guide this discussion are provided in Mandalios (2013) and ESM-A.

The instructor should then ask groups to divide the responsibilities for evaluating the video (using the RADAR components) among group members so as to not overwhelm them. By focusing on one, two, or three criteria/steps, each student should be able to critically evaluate the video while also understanding the content. If the instructor plans to implement part 2 (see below), students should also be instructed to make note of any "factual" claims that seem outrageous or biased and that they will be asked to further investigate and report on their findings in a follow-up assignment. The instructor should then play the "Story of Stuff" video (Leonard 2007) for the class (22 min), during which each student should take notes in general and about their RADAR component. At the end of the video, allow the groups 5–10 min to share and prepare for class dialogue by developing a list of at least two strengths and two weaknesses of the "Story of Stuff" based on the RADAR criteria. Then have each group provide one strength and one weakness of the video (as a reliable source of information) and record these in a public location (whiteboard, chalkboard, overhead projector) for a collective class critique (10–15 min). The collaborative nature of this portion of the activity typically results in a rigorous student-driven list of reasons that the "Story of Stuff" video is valuable as an advocacy video but is not a fully reliable source of information (see ESM-B for examples of critique comments). If the instructor chooses to proceed with part 2 of the activity, each group should be asked to choose one "factoid" or claim to research and then be given the homework research assignment and expectations for part 2. If time permits, concluding discussion could be facilitated using the prompts suggested in the follow-up section (if not to be used to end part 2).

Part 2: Determining Validity of a Claim and Research Source Reliability

To reinforce the use of RADAR, it is suggested that the instructor require students to list both reliable and unreliable web-based sources to show the class how they determined the credibility (or lack thereof) of each. Depending on the academic level of the students, the instructor may require students to include peer-reviewed sources in the research; however, the main objective of the assignment is to encourage students to critically evaluate sources that appear in a normal Internet search, some of which may be scholarly. Presentation of student findings should highlight the subject of investigation (the "factoid" or "claim") conclusions based on their research and a summary of their reliable and unreliable sources (see ESM-B for suggestions). Details should be based on instructor preference, assessment needs, and available time. If desired, extended collaborative engagement could be facilitated using the discussion questions listed in the next section or others relating to objectives of the course in general.

Follow-Up Engagement

- Suggested discussion questions include:
    - What is the best way to find information about a topic of interest?
    - Why is the date of development of the source important? Are there situations under which it might be more important than others?
    - What are potential motivations of resource developers, and what indicators might one use to help determine motivations?
    - Do you think that the general public, in general, analyze the sources they use for information?
    - For what reason(s) would it be important to use the most credible sources to develop opinions? Are there risks associated with trusting biased sources or sources that may not be dependable? What would those be?
- Students can be provided with another opportunity to investigate the web for unreliable sources. For instance, the tendency to assume that ".org" websites are reliable sources of information can be combated with an assignment that requires students to locate and document several reliable and unreliable ".org" (or .com, etc.) sources on a sustainability or environmental topic. A suggested work sheet and examples are provided in ESM-B.
- Students' skills in evaluating the reliability of sources may be assessed by providing several sources and having students evaluate their reliability or by having students find examples of reliable and unreliable sources on a given topic. Examples are provided in ESM-B. (Also see Chap. 36, this volume, for an alternative approach.)

Connections

- This lesson can enrich discussions regarding decisions in environmental policy development, politicization of sustainability development issues, and public response.

- Any activity involving research or debate can be used as an opportunity to reinforce the critical evaluation skills developed in this activity. These skills will promote the use of higher levels of thinking to investigate and discuss controversial issues including (but not limited to) climate change, energy sources and use (Chap. 26, this volume), biodiversity (Chap. 18), sustainable development, triple bottom line considerations (Chaps. 23 and 24), consumerism (Chap. 22), environmental toxins (Chaps. 31–33), and many others.

References

Doren L (2009) Story of Stuff, the critique. https://www.youtube.com/watch?v=c5uJgG05xUY. Accessed 1 Mar 2014

Kaufman L (2009) A cautionary video about America's 'stuff'. http://www.nytimes.com/2009/05/11/education/11stuff.html. Accessed 10 Jan 2015

Leonard A (2007) The Story of Stuff. http://storyofstuff.org/movies/story-of-stuff/. Accessed 1 Aug 2009

Mandalios J (2013) RADAR: an approach for helping students evaluate Internet sources. J Inf Sci 39:470–478

Mark (2009) The Story of Stuff critique—a critique. http://breathenetwork.org/2009/10/08/story-of-stuff-critique-%E2%80%93-a-critique/. Accessed 1 Aug 2009

McClure R, Clink K (2009) How do you know that? An investigation of student research practices in the digital age. Portal Libr Acad 9:115–132

Metzger MJ, Flanagin AJ, Zwarun L (2003) College student Web use, perceptions of information credibility, and verification behavior. Comput Educ 41:271–290

Metzger MJ, Flanagin AJ, Medders RB (2010) Social and heuristic approaches to credibility evaluation online. J Commun 60:413–439

Stapleton P (2005) Evaluating web-sources: internet literacy and L2 academic writing. ELT J 59:135–143

White R (2013) Beliefs and biases in web search. In: Proceedings of the 36th international ACM SIGIR conference on research and development in information retrieval, Dublin, Ireland, July 28–August 01, 2013

Chapter 36
Critically Evaluating Non-Scholarly Sources Through Team-Based Learning

Nathan Ruhl

Introduction

Non-scholarly sources of information (e.g., non-peer-reviewed popular press and mass media articles) have been shown to shape individual opinions about global climate change (GCC) and subsequently isolate individuals from dissenting viewpoints via a "reinforcing spiral" (Feldman et al. 2014; Chap. 39, this volume). The diversity of non-scholarly material surrounding the public GCC debate provides an opportunity to critically evaluate these sources and demonstrate the dangers of citing non-scholarly sources. There is strong evidence that students' citation behavior can be redirected toward scholarly sources via the implementation of penalties (Davis 2003; Robinson and Schlegl 2004, 2005), but this approach does not teach students why they need to use scholarly sources. Alternative teaching strategies (peer teaching, cooperative learning groups, games, etc.) may increase the use of scholarly sources and result in increased awareness of the importance of citing them. For instance, in a recent example, Markey et al. (2012) showed that the use of a game can increase the quality of student citations. Through the use of a modified team-based learning (TBL) activity (Box 36.1), the goal of the exercise presented here is to use non-scholarly reports about GCC as a tool for students to discover that non-scholarly sources must be critically evaluated, especially when the topic may be controversial to the public (also see Chap. 35, this volume). In this activity students will be assigned readings as homework, answer questions about the readings individually and with a small team, and then work with their teams to understand broader issues raised by the readings.

Electronic supplementary materials: The online version of this chapter (doi:10.1007/978-3-319-28543-6_36) contains supplementary material, which is available to authorized users.

N. Ruhl (✉)
Department of Biological Science, Rowan University, Glassboro, NJ, USA
e-mail: Ruhl@Rowan.edu

© Springer International Publishing Switzerland 2016
L.B. Byrne (ed.), *Learner-Centered Teaching Activities for Environmental and Sustainability Studies*, DOI 10.1007/978-3-319-28543-6_36

Learning Outcomes

After completing this activity, students should be able to:

- Know that non-scholarly sources need to be critically evaluated because they may not be reliable.
- Recognize that all sources of information are potentially biased.
- Discuss, in a general sense, how and why GCC science can be misreported and/or misconstrued by non-scholarly sources.

Course Context

- Originally designed for an upper-level undergraduate biology course on global climate change
- 90 min in one class period but easily adaptable to shorter meetings
- Used with three to ten teams of three to ten students each (for a range of 9–100 total students)
- Students complete readings at home but no other student background is necessary
- All components are non-technical; high school and undergraduate students of any major should be able to engage with the activity

Instructor Preparation and Materials

Before the activity, the instructor should become familiar with the current state of global climate change research. An excellent summary is provided by the "What We Know" website hosted by the American Association for the Advancement of Science (AAAS 2014). The instructor should then review the set of seven readings that will be assigned to students (mass media articles; ESM-A). The readings have been chosen to reinforce that critical reading of non-scholarly sources is essential and all sources of information are not equal in terms of their reliability and bias. To maintain "timeliness," the instructor may want to update/alter the readings/questions as the state of GCC research and the public debate about it progresses. When making updates, ensure that the readings/questions support the discussion points listed at the end of the section "Activities." Also review the six iRAT/gRAT questions and five application questions (defined in Box 36.1), editing them as needed to meet the course context and needs (ESM-B). Answers and related discussion points for these questions are given in ESM-C. The iRAT/gRAT questions are not directly related to the goals and learning outcomes of the activity per se; they are designed to ensure that students have carefully read the assigned material.

The activity requires that the students be placed into at least three teams of at least three students each and that the teams be as diverse as possible in terms of demographics and student backgrounds (based on recommended practices for

Box 36.1. Team-Based Learning (TBL)

Team-based learning (TBL) is a flexible form of cooperative/group learning where students are assigned to a team (group) and then work with that team to solve complex problems or answer complex questions (Michaelsen et al. 1982). The teams need to be large and diverse enough (in terms of gender, age, background, race, etc.) that multiple points of view will be represented. The groups are referred to as teams because there can be an element of shared grades and because TBL activities can be chained together throughout a course with the teams maintained. An element of competition is naturally introduced by the formation of teams such that teams may try to "win" the day's activity by answering the most questions correctly. This behavior is desirable because it reinforces learning outcomes; the "winning" team is the one that has done the best collective job of learning the assigned material. However, the instructor should be careful about acknowledging "winners" so as not to demotivate those that did not "win." A typical TBL class activity involves an individual readiness assessment test (iRAT), group readiness assessment test (gRAT), and a series of application questions (all of which can be graded). The goal of the iRAT is to ensure that individual students read and understand the assigned material. The goal of the gRAT is for the members of the team to come together with their answers from the iRAT, discuss why they answered the way they did, and (hopefully) come to the correct answer while at the same time reinforcing the usefulness of their teammates in determining the correct answer. Because the same questions are used for the iRAT and gRAT, the questions are intentionally difficult. The application questions are where the "real" learning is intended to occur. Students should work through the application questions one at a time. Once all teams have finished a question, the instructor then leads a short discussion with the whole class about which answer is the correct one and why. A good way to do this is to allow the students to explain why they answered the way they did before revealing the correct answer and explaining why it is correct. More information about group learning techniques can be found in Davidson et al. (2014). More information about TBL (including best practices for writing TBL questions) can be found in Michaelsen et al. (2002) and at http://www.teambasedlearning.org.

TBL; see Box 36.1). The instructor will need to print a sufficient number of the iRAT/gRAT questions (one set for each student and at least one set per team). The application questions could be printed out (one set per group), provided electronically, or displayed with a projector, but it is recommended that each question be provided one at a time given the structure of how they are to be discussed.

The activity, as outlined below, should take 75–90 min. It could be shortened by eliminating one or both of the iRAT/gRAT components (which each take about 15 min) and/or the instructor could simply lead a short class discussion of the application questions without breaking the class into teams. Alternatively, the activity could be lengthened to any desired time through the addition of iRAT/gRAT or application questions or extending discussion.

Activities

Students should have completed the reading assignments as homework. In class, students are given 10–15 min to individually answer the iRAT, their answers are collected, and they are placed into predetermined teams to collaboratively complete the gRAT (15–20 min). The remaining time is used for working through the application questions in teams and integrated whole-class discussion. It is important that the application questions be worked through one at a time. Each team discusses a question, comes to a consensus on the correct answer, and then waits for the other teams to do the same. While this may leave teams with some "downtime" as they wait, in practice it often leaves time for the team to rethink their answer as they overhear the discussion other teams are having around them, and discussion may be rekindled (though "espionage" should not be encouraged). If a team comes to a conclusion too quickly, the instructor could challenge the team's conclusion (e.g., play devil's advocate) or ask follow-up questions.

After all teams have finalized their answer for a question, the instructor asks each team to reveal their answer to the rest of the class. If there are a large number of teams, the instructor may want to tally the answers on a board to visualize the level of agreement between teams. If there is disagreement among teams, each should be asked to defend its answer and why they think that answer is correct before the instructor reveals the correct answer and leads a relevant discussion (talking points to guide this discussion are provided in ESM-C and summarized below). The class then moves on to the other questions in turn. The discussion of each application question (1–5) should highlight at a minimum: (1) how special interest groups, lobbyists, and politicians spin the facts about GCC toward whatever conclusion is most convenient or profitable for them; (2) that the best method to combat bias or conflicts of interest is to acknowledge those issues, not artificially "balance" bias by introducing opposing viewpoints; (3) motivations for misrepresentation of content; (4) mistakes vs. misrepresentation; and (5) the authority and reliability of different types of information relative to the peer-reviewed primary literature (see ESM-C for more detail).

Follow-Up Engagement

- Ask students to read a peer-reviewed paper and then read a non-scholarly summary of that paper. Compare/contrast how the authors of both works present the topic (see examples in Chaps. 14 and 15, this volume).

- Assign a climate change myth that is perpetuated by non-scholarly sources and either write a paragraph about why it is a myth or give a short presentation to the class. An extensive list of climate change myths (and scientific explanations for why they are myths) is provided by Cook (2015).
- Provide students with a list of online readings and ask them to rank them on scales of authoritative and non-authoritative, biased and non-biased, or scholarly and non-scholarly. Discuss how authority, bias, and scholarship are interlinked and the influence of these factors on the necessity for critical reading.

Connections

- Discuss the reliability of the non-scholarly literature for providing information about other controversial topics (e.g., biodiversity conservation, genetically modified organisms).
- Recall this lesson when referencing online material and news in discussions of other topics.

Acknowledgments Thanks go to my climate change students at Rowan University for helping develop this activity and to R. Hoffman, O. Lopez and K. Behling for introducing me to TBL.

References

AAAS (2014) What we know. http://whatweknow.aaas.org/. Accessed 1 Mar 2015

Cook J (2015) Global warming and climate change myths. https://www.skepticalscience.com/argument.php. Accessed 1 Mar 2015

Davidson N, Major CH, Michaelsen LK (2014) Small-group learning in higher education—cooperative, collaborative, problem-based, and team-based learning: an introduction by the guest editors. J Excel Coll Teach 25:1–6

Davis PM (2003) Effect of the web on undergraduate citation behavior: guiding student scholarship in a networked age. Portal Libr Acad 3:41–51

Feldman L, Myers TA, Hmielowski JD et al (2014) The mutual reinforcement of media selectivity and effects: testing the reinforcing spirals framework in the context of global warming. J Comm 64:590–611

Markey K, Leeder C, Taylor CL (2012) Playing games to improve the quality of the sources students cite in their papers. Ref User Serv Q 52:123–135

Michaelsen LK, Watson WE, Cragin JP et al (1982) Team-based learning: a potential solution to the problems of large classes. Exchange Organ Behav Teach J 7:13–22

Michaelsen LK, Bauman-Knight A, Fink LD (2002) Team-based learning: a transformative use of small groups. Praeger, New York

Robinson AM, Schlegl K (2004) Student bibliographies improve when professors provide enforceable guidelines for citations. Portal Libr Acad 4:275–290

Robinson AM, Schlegl K (2005) Student use of the Internet for research projects: a problem? Our problem? What can we do about it? PS 38:311–315

Chapter 37
Using the Insight, Question, and Challenge (IQC) Framework to Improve Students' Environmental Communication Skills

Seaton Tarrant

Introduction

Environmental issues are complex because indeterminacy is part of the emergent character of many social and ecological systems. Decisions regarding environmental sustainability are increasingly negotiated through participatory, democratic deliberations, where scientific facts are considered alongside a plurality of normative opinions regarding acceptable trade-offs and quality-of-life standards (Lee 1994; Williams and Matheny 1998; Prugh et al. 1999). Thus, decision-making at the policy-science-sustainability interface requires that scientists, policy makers, and citizens keep an open mind regarding others' viewpoints while also effectively, critically, collaboratively, and civilly evaluating and discussing scientific evidence and personal positions.

The skills needed to achieve democratic and sustainable solutions can be cultivated in the sustainability classroom through activities that allow students to practice and improve effective communication through a structured dialogue (Vella 2002; Tillbury 2011; Wiek et al. 2011; Wals 2012). Such activities foster increased social interactions, which can improve cognitive elaboration abilities (Frijters et al. 2008). It also gives students the opportunity to guide the overall flow of the class

Electronic supplementary materials: The online version of this chapter (doi:10.1007/978-3-319-28543-6_37) contains supplementary material, which is available to authorized users.

S. Tarrant (✉)
Sustainability Studies, University of Florida, Gainesville, FL, USA
e-mail: seatonius@ufl.edu

© Springer International Publishing Switzerland 2016
L.B. Byrne (ed.), *Learner-Centered Teaching Activities for Environmental and Sustainability Studies*, DOI 10.1007/978-3-319-28543-6_37

discussion based on what they have found most insightful and intriguing within readings.

The goal of the learning activity described below is to help students critically and constructively read, reflect on, and discuss complex environmental and sustainability issues through practice. This is facilitated through the use of the insight, question, and challenge (IQC) framework, which helps students through an iterative process of careful reading, note-taking, and peer-to-peer interaction. For each reading, they will record their insights, questions, and challenges and use these to guide a dialogue with a classmate who will practice listening and taking notes. The activity can be used to explore a range of environmental and sustainability topics using news articles to engage with current events and peer-reviewed literature to expose students to research. Further, as described in the "Follow-Up Engagement" section, the IQC activity can be modified in several ways, allowing for multiple implementations in one course, thus providing iterative opportunities for students to practice and improve their skills.

Learning Outcomes

After completing this activity, students should be able to:

- Critically challenge a given reading for its quality of information, regarding both facts and values.
- Reflect on environmental and sustainability issues to generate and communicate personal insights.
- Listen carefully and critically to, and record pertinent information about, a peer's ideas on a shared reading.
- Synthesize their own and a peer's perspectives about a reading.

Course Context

- Developed for an introductory-level class on sustainability and used in an intermediate-level course on politics and sustainability and an advanced environmental ethics and politics course
- Has been implemented with students of all class levels, from environmental science majors to business majors, and with as few as 6 and as many as 50 students
- 20–45 min in one class meeting
- Before class, one or two readings should be read by the students
- Adaptable to any course and class size that includes contested environmental and sustainability issues

Instructor Preparation and Materials

To complete this activity with students, the instructor should choose one or two readings that fit the context of the course. They can be short readings about current events, such as a newspaper or magazine article or peer-reviewed articles. Importantly, each reading should include empirical facts about the issue under discussion. It is helpful if the value position of the author is also explicit, though, if implicit, this can be derived as part of the reading exercise. Readings do not need to be "perfect" in their presentation of facts or stated normative bias; uncovering these shortcomings is part of the experience of critically engaging with the text. A list of articles that have been used for the activity is included in Electronic Supplementary Materials (ESM)-A. Readings should be assigned to and completed by the students prior to the activity.

The IQC worksheets (ESM-B) include both a reading and a listening worksheet. Explanations for each aspect of the IQC can be found in the section "Activities" below. Instructors should review the worksheets, adjust them if necessary to fit their class context, and print one reading and one listening worksheet for each student (both will fit on a single double-sided sheet of paper). In order to introduce students to the worksheet, instructors may wish to prepare a completed IQC worksheet example for one of the chosen readings or from an unassigned reading, depending upon whether the instructor would like the students to complete IQC's for one or two readings during the first iteration. Instructors can also choose to use the completed IQC that is included in ESM-A. During this preparation time, the instructor might also consider looking through the supplementary materials provided in ESM-C, which connect the activity with broader inquiries in sustainable resource management and environmental deliberation.

As desired, prepare a bowl or hat (a silly one, like a clown's, can create a more fun atmosphere) that contains the name of each student on a small slip of paper that can be used to identify students who will share their responses with the whole class. Also, before the day of the activity, decide whether students will fill out their first IQC on their own or in small groups; in the author's experience, sometimes, students find initial success filling out the IQC in groups of two to four.

Activities

Distribute the IQC reading and listening worksheets and review their components (5 min). Explain to the students that "insights" involve creative thinking that connects the readings to their own lives, generating relevance and meaning. When reviewing the "questions" section, stress that students can use this section to expand the conversation beyond the obvious points or to address parts of the article that are unclear. For the "challenges" section, explain that thinking critically doesn't necessarily mean rejecting the points in a reading; stress that critical reading is an important part of clarifying facts and understanding arguments, and suggest that there is always room for improving or advancing a line of inquiry.

Next, if one has been prepared, present the completed IQC that introduces students to the worksheet. Project the completed form on a large screen or read it aloud to the class. The instructor can help identify the debatable points from the reading and stress that such complexity is characteristic of most environmental situations and that communication and analytical skills are a key part of developing sustainable solutions. The instructor should stress that the exercise is practice to this end.

After this introduction, have students individually complete the IQC in class for another reading (15–20 min). The instructor should walk around the room, fielding any questions about the exercise, and motivating students to surprise themselves with unique interpretations and strong critical thinking. Once students have completed the IQC, ask them to pair up and share their responses with a partner. At this point, each student will fill out the IQC listening worksheet while listening to a partner's presentation of his or her ideas. Explain that they each have 5 min to share and that they should try to take notes on their partner's ideas quickly and efficiently. Emphasize that this is practice in speaking and listening, so students shouldn't simply pass each other's IQC forms to each other and copy their partner's answers.

Alternatively, the instructor may choose to initially present students with only the IQC reading worksheet. If this strategy is taken, the instructor does not pass out the listening worksheet until after the students have shared their reading worksheet responses with a peer. Thus, the students complete their listening worksheet, as described in the previous paragraph, from memory. This approach has proven successful at bringing students' attention to the fact that their listening skills have room for improvement, garnering more student interest in and reflection on the activity.

After students have finished the listening worksheet, engage the whole class in open discussion, giving students an opportunity to reflect on what they learned from their peers. (As desired, students' names can be drawn from a hat to facilitate this.) Prompting students to consider the following can spark discussion:

- Are multiple interpretations of the reading possible?
- How do we decide which interpretation is more "correct"?
- What does it mean to critically evaluate a reading and does the IQC form help this?
- Has your own perspective been challenged or reinforced by the exercise?

Follow-Up Engagement

- In future iterations, have the students fill out the IQC reading worksheet as part of their homework, and come to class prepared to complete the IQC listening worksheet with a partner. The activity can be used as a weekly or bi-weekly exercise, with students keeping a notebook of their IQC's. Student journals can be assigned in which they reflect on and evaluate how their critical reading responses and listening skills are changing over time.

- The amount of student-student interaction can be increased by extending the activity for a longer period. This can be done with "speed sharing," in which students move from peer to peer in short intervals (2 or 3 min), keeping notes and quickly collecting a larger variety of perspectives on an issue, which can subsequently be summarized and synthesized.
- Have students write their own letter to the editor or op-ed article (Chap. 40, this volume) on an important environmental issue, and then have students fill out IQC's on each other's articles.
- Plan a class debate on a contentious issue, such as GMO's, fracking, biodiversity conservation (Chap. 18, this volume) or how best to mitigate climate change, and have students prepare for the debate using the IQC framework. (Also see Chaps. 29 and 39, this volume, for examples of structured discussion approaches.)

Connections

The activity can be aligned with:

- Advanced discussions regarding the potential for democratic solutions to address complex environmental problems as in mock town hall or UN debates.
- Introductions to adaptive and collaborative resource management in which the roles and perspectives of various stakeholders (e.g., scientists and citizens) need to be considered and negotiated to develop environmental policy (e.g., see Chap. 19, this volume).
- An introduction to media literacy, in which students identify and negotiate media bias and green washing in popular media (e.g., see Chaps. 34–36, this volume).

References

Frijters S, ten Dam G, Rijlaarsdam G (2008) Effects of dialogic learning on value-loaded critical thinking. Learn Instruct 18:66–82

Lee KN (1994) Compass and gyroscope: integrating science and politics for the environment, 2nd edn. Island Press, Washington, DC

Prugh T, Costanza R, Daly HE (1999) The local politics of global sustainability, 1st edn. Island Press, Washington, DC

Tillbury D (2011) Education for sustainable development: an expert review of processes and learning. DESD monitoring and evaluation. UNESCO, Paris

Vella JL (2002) Learning to listen, learning to teach: the power of dialogue in educating adults. Jossey Bass, San Francisco, CA

Wals AE (2012) Shaping the education of tomorrow: 2012 full-length report on the UN decade of education for sustainable development. DESD monitoring and evaluation. UNESCO, Paris

Wiek A, Withycombe L, Redman CL (2011) Key competencies in sustainability: a reference framework for academic program development. Sustain Sci 6:203–218

Williams BA, Matheny AR (1998) Democracy, dialogue, and environmental disputes: the contested languages of social regulation. Yale University Press, New Haven, CT

Chapter 38
Building Students' Communication Skills and Understanding of Environmental and Sustainability Issues Interactively and Cumulatively with Pecha Kucha Presentations

Bonnie McBain and Liam Phelan

Introduction

This chapter describes the use of Pecha Kucha (Japanese for the sound of conversation or "chitchat") as a presentation format to facilitate class discussions. Pecha Kucha is a highly constrained presentation format requiring exactly 20 simple, image-based slides with minimal or no text that are each presented for 20 s (for a total presentation length of 6 min and 40 s) (Pecha Kucha 2014). The twofold goal of using this approach as a learning activity is to build students' (i) understanding of, and (ii) communication skills about, environmental and sustainability issues (Caron and Serrell 2009; Phelan et al. 2015; Rieckmann 2012).

Preparing for the presentation is pedagogically valuable because it helps students to (i) research an issue and (ii) use critical thinking to extract its essence. The process of having to articulate knowledge very briefly results in deeper understanding because it requires iterative refinement of thought (Herrington and Herrington

Electronic supplementary materials: The online version of this chapter (doi:10.1007/978-3-319-28543-6_38) contains supplementary material, which is available to authorized users.

B. McBain (✉)
School of Environmental and Life Sciences, University of Newcastle,
Callaghan, NSW, Australia
e-mail: Bonnie.McBain@newcastle.edu.au

L. Phelan
School of Environmental and Life Sciences, University of Newcastle,
Callaghan, NSW, Australia

Zanvyl Krieger School of Arts and Sciences, Johns Hopkins University,
Washington, DC, USA
e-mail: Liam.Phelan@newcastle.edu.au

© Springer International Publishing Switzerland 2016
L.B. Byrne (ed.), *Learner-Centered Teaching Activities for Environmental and Sustainability Studies*, DOI 10.1007/978-3-319-28543-6_38

Box 38.1. Using Pecha Kucha Presentations in Online and Hybrid Courses

Pecha Kucha presentations can be pre-recorded asynchronously for online classes. In hybrid contexts, e.g., flipped classrooms, students can view pre-recorded presentations outside of class time and use class time for discussion. In fact, this activity was originally developed for use in online courses. More technical information on how to record Pecha Kucha presentations and then upload them to a course's discussion board is available in ESM-B. In the online environment, the instructor may need to moderate online discussions and provide additional feedback about appropriate etiquette specifically for online conversations (e.g., toning down language to manage the tone of conversations, paying particular attention to clarity, and being brief and to the point).

2006). Further, giving the presentation requires students to (iii) communicate effectively through concise and dynamic presentations and subsequent class discussions, skills that are essential for graduates from environmental and sustainability studies programs (Phelan et al. 2015; Rieckmann 2012). Finally, (iv) presenter-generated questions serve as prompts for subsequent discussions to encourage critical thinking and collaborative learning by both the student facilitator and her peers.

The Pecha Kucha format is appropriate for multiple topics across environmental and sustainability studies and can be used for in-class, live presentations or adapted for online or hybrid courses (see Box 38.1). It is also suitable for contexts in which students research and present about topics they choose or that are assigned by the instructor. Pecha Kucha presentations by all students can be combined strategically to cover different aspects of an entire course, or parts of it, to support students' cumulative, holistic learning, in a manner similar to jigsaw and other cooperative learning strategies (see SERC 2015). A further benefit of this approach is that a lot can be learned by students in a short time due to the presentations' short duration.

Learning Outcomes

After presenting and engaging in discussion with peers, students should be able to:

- Research an environmental issue and use critical thinking to distil its salient, core elements.
- Communicate effectively about environmental issues through Pecha Kucha presentations and asking related open-ended questions.
- Develop a broader understanding of environmental issues through watching and discussing peers' presentations.

Course Context

- Originally designed for Masters-level environmental courses in which students were allocated into tutorial groups of 10–20 and where the majority of weekly classes are given over to student presentations. The majority of class time each week was dedicated to student-facilitated discussion using presentations and discussion questions
- Each student presentation takes 6 min and 40 s
- Student preparation includes research, writing a well-planned presentation script, preparing images for the presentation, practicing delivery, or recording their presentation. Students also write open-ended questions which are subsequently used in presenter-facilitated tutorial discussions with peers following the presentation
- Adaptable to many class sizes, topical contexts, and use for online courses

Instructor Preparation and Materials

In the first weeks of the semester, the instructor schedules student presentations in one of two ways: (1) allocating students to topics or (2) inviting students to choose presentation topics which fit strategically with the course's learning goals. Once topics and presentation dates are set, the instructor can allocate presentation research prompts (questions) if they want students to focus on particular aspects of a topic. This is useful if a group of students are presenting the same topic in the same class as it ensures that different aspects of each topic are explored by each student. For example, one topic in the authors' *Environmental Management* course was "Environmental Governance," and an assigned research prompt was "Discuss the strengths and limitations of deliberative democracy as an approach to environmental management. Use an example to demonstrate." Prescribing this prompt ensured that this particular aspect of environmental complexity was specifically explored as a critical part of managing environmental complexity as a whole. Asking students to choose their own case study both ensured the practical relevance of their learning and encouraged them to customize their presentation to something that interested them.

The highly constrained Pecha Kucha format—exactly 20 s of narration per slide, exactly 20 slides, and an emphasis on images—forces students to distil the salient elements of what it is they want to communicate to their peers. The instructor can provide guidance to students about how to do this through tips for how to find relevant research using library or other online databases. As desired, the instructor can also provide guidance on effective communication about environmental and sustainability issues based on information provided by Cox (2013) and Harding et al. (2009). Sharing a grading rubric with students before they present can help communicate to them what they should be aiming for (see Electronic Supplementary Materials

(ESM)-A for an example rubric). If time is available, the instructor can offer formative feedback on students' draft presentation transcripts and discussion questions *before* the students present to peers. This ensures that both the instructor and presenter are confident in the quality of the material being presented. As desired, the instructor can also help build students' confidence in making these very structured presentations by providing encouragement and guidance, depending on the students' academic background and experiences.

Although the post-presentation discussions should be facilitated by the students, the instructor retains responsibility for facilitating the overall process to ensure that students understand the guidelines and etiquette for communicating effectively and respectfully with one another. This includes making clarifications to student responses where required, encouraging thoughtful contributions, and asking follow-on questions that deepen student engagement with the material.

Finally, the role of the instructor is to help students see how each of the presentations and discussions "fits" into the bigger picture of the learning goals of the course. This can be done by providing students with an initial introduction to the overall topic and then introducing each presentation topic individually by highlighting how that contributes to their overall learning. The instructor can also link the topics of previous presentations with those that are currently being presented by contributing to student discussions.

Activities

Preparation time for students is substantial; the tasks require students to carefully research and then shape their presentation with suitable images and a well-planned script and then practice delivering their presentation. Students should be given clear direction about two aspects of the Pecha Kucha format (see Shenk 2009). First, students need to understand the Pecha Kucha is a precisely constrained and specific form of presentation. It is important to emphasize to students that they do not try to present too much information (see ESM-B for a guide on how to show students to calculate how much content will comfortably fit into a 20-s slide). Modeling a high-quality Pecha Kucha or providing links to online examples can be very helpful in this regard. Second, students can benefit from clear step-by-step guidance on the technical steps to creating a Pecha Kucha. Students can be encouraged to test the presentation software they will use before presenting so that they are confident that technical issues will not delay them during their presentations. Instructors could also provide to students a template file with all the settings already in place. More detailed instructional guidance on all the above aspects of teaching Pecha Kucha can be found in ESM-B.

Students also write open-ended questions for presenter-facilitated tutorial discussions following the presentation. Open-ended discussion questions require more than a "yes" or "no" answer and are useful in establishing and maintaining thoughtful contributions to tutorial discussions. Examples include: What can different disciplinary perspectives offer in understanding environmental controversies? Identify diverse stakeholders in an environmental controversy, and describe how their perspectives on the controversy align and/or contrast.

In some contexts, such as with less academically mature cohorts, students may also benefit from guidance on how to constructively engage in tutorial discussions. For example, students can be encouraged to discuss presentations from a variety of angles and with open minds and to introduce potentially unrepresented perspectives. They can also be encouraged to listen and deliberate on the perspectives raised by other students. Suggesting to students that they do not necessarily need mastery over the subject matter to engage in discussions can alleviate nervousness that some students may have. For instance, inviting students to share their understanding of a previous point can help clarify their own learning (by having to articulate it) while opening the possibility of asking for clarification if it is required.

As facilitators of discussions, students can be guided in several ways. They can be assured that their role is to open up an area for learning with their peers and that they do not need to have all the answers. Rather they can ask follow-on questions based on the responses of other students and ask the instructor for clarification. When peers express diverse, alternative opinions, they can be encouraged to be respectful with responses such as: "I wonder if....," or "Could I play devil's advocate....,"' or "That is a great point you make but what if ...." Most importantly, students can be encouraged to be welcoming and inclusive to facilitate engaging, productive discussions.

Follow-Up Engagement

- Peer-to-peer evaluation of presentations can be incorporated into the presentation sessions. This encourages students to reflect upon the characteristics of the material presented, the effectiveness of the communication, and the attributes of good discussion techniques. Providing students with a clear rubric ahead of time (e.g., ESM-A) that defines effective skills in these areas will provide a guide to how they can assess their peers' work.
- Students can be asked to write a reflective journal which documents their realizations and experiences from interacting with the Pecha Kucha presentations. This can be submitted as an assessment task itself, or students can summarize their reflections further and submit that as their assessment task (thus requiring "reflection on reflections").

Connections

- If the presentations are scheduled over several course meetings and/or weeks, the instructor and students can refer back to previous presentations, identifying ways in which constitutive elements of an overall curriculum connect to each other.

This supports students' cumulative, holistic learning and demonstrates to students the tangible value of their work for informing ongoing learning.

- Pecha Kucha presentations could be recorded to transform them into artifacts which can have longer-term benefits. They can be revisited later in a course to review material based on subsequent learning. In addition, after individual courses have concluded, students can include their presentations in learning portfolios.

References

Caron RM, Serrell N (2009) Community ecology and capacity: keys to progressing the environmental communication of wicked problems. Appl Environ Educ Commun 8:195–203

Cox R (2013) Environmental communication and the public sphere, 3rd edn. Sage, Thousand Oaks

Harding R, Hendricks C, Faruqi M (2009) Environmental decision-making—exploring complexity and context. The Federation Press, Sydney

Herrington A, Herrington J (2006) What is an authentic learning environment? Information Science Publishing, London

Leach M, Scoones I, Stirling A (2010) Dynamic sustainabilities—technology, environment and social justice. Earthscan, London

Rieckmann M (2012) Future-oriented higher education: which key competencies should be fostered through university teaching and learning? Futures 44:127–135

Pecha Kucha (2014) Pecha Kucha. www.pecha-kucha.org. Accessed 17 Oct 2014

Phelan L, McBain B, Ferguson A et al. (2015) Environment and sustainability learning and teaching academic standards statement. Office for Learning and Teaching. environmentltas. gradschool.edu.au/. Accessed 21 July 2015

Science Education Resource Center (SERC) (2015) Pedagogy in action—jigsaws. http://serc.carleton.edu/sp/library/jigsaws/index.html. Accessed 6 May 2015

Shenk M (2009) The case for complexity, the Pecha Kucha way. http://www.anecdote.com/2009/07/the-case-for-complexity-pecha-kucha-way/. Accessed 26 Feb 2015

Chapter 39
Engaging in Climate Change Conversations: A Role-Playing Exercise to Cultivate Effective Communication

Stephen Siperstein

Introduction

Public understanding of climate change in the USA is fragmented across political and ideological boundaries and has become more a question of belief than of understanding (Leiserowitz et al. 2012). Moreover, research indicates that the climate change "debate" is largely a function of differences in individual ideology and thus best understood as "an intergroup conflict that exists primarily between two groups with conflicting views…believers and skeptics (rather than between scientists and sections of the public)" (Bliuc et al. 2015, p. 226). Among the many barriers to effective climate change communication—such as media selectivity (Feldman et al. 2014) and the public's skewed perception of scientific agreement (Ding et al. 2011)—this ideological polarization is central. Effective climate change communication must therefore do more than attempt to convey information or knowledge (Moser and Dilling 2011, p. 163); it needs to facilitate conversation, connection, and perspective taking, so that individuals can begin to talk *with* (rather than *at* or *past*) each other about climate change (Hobson and Niemeyer 2012; also see Chaps. 4, 29 and 40, this volume). Unfortunately, most people rarely talk about climate change with each other, and when they do, they usually have trouble with perspective taking and empathy. Marshall (2014) specifically identifies this lack of positive and productive climate change conversations as part of a pervasive culture-wide "meta-silence" about the issue ("meta" because people do not talk about not talking about it).

Electronic supplementary materials: The online version of this chapter (doi:10.1007/978-3-319-28543-6_39) contains supplementary material, which is available to authorized users.

S. Siperstein (✉)
Department of English, University of Oregon, Eugene, OR, USA
e-mail: stephen.siperstein@gmail.com

© Springer International Publishing Switzerland 2016
L.B. Byrne (ed.), *Learner-Centered Teaching Activities for Environmental and Sustainability Studies*, DOI 10.1007/978-3-319-28543-6_39

Box 39.1. Role-Play and Climate Change Education

Role-play is a creative, participatory, and potentially transformative type of learning activity. In the phrase itself, the word "role" suggests that students gain experience with perspective taking (and potentially with empathy), and the word "play" indicates that students use their imaginations to have fun and act out different parts in a supportive environment. Within the context of teaching climate change, role-play can provide a framework for students to explore the issue's various intellectual and emotional dimensions. Additionally, the climate change role-playing activity described here invites students to reflect on the act of conversation itself.

The "meta-silence" on climate change suggests important and pressing questions for environmental educators, such as how best to instruct students about the challenges to effective communication and how to help them develop the skills and confidence necessary to have productive conversations about climate change. Student-centered teaching activities that use role-play can be one avenue for tackling such questions (see Box 39.1). The goal of the activity described in this chapter is to help students recognize the various challenges of climate change conversations and to experience them through imaginative role-play. To achieve this goal, students consider how others might view climate change—based on ideological positions, worldviews, religious beliefs, professions, life stories, etc.—and then "play" characters whose perspectives on climate change may be different from their own. By viscerally experiencing the challenges of having such conversations, students should become better able to assess and overcome the barriers to effective communication about complex, controversial environmental issues such as climate change.

Learning Outcomes

After completing this activity, students should be able to:

- Reflect on their own worldviews and perspectives about climate change.
- Investigate a particular perspective on climate change that may differ from their own.
- Explain why communication about climate change can be challenging.
- Identify strategies for having positive and productive climate change conversations.
- Consider why having conversations about climate change can be important for addressing the ideological barriers to effective climate change communication.

Course Context

- Developed for a first-year seminar focused on climate change with 15–20 students
- 80 min in one class meeting

- Students should have a basic knowledge of climate change and read or otherwise engage with various sources on climate change communication before class.
- Adaptable to courses of any class size and level that include discussions of climate change and communication; it can be modified for longer or shorter durations by assigning additional preparatory work outside of class or by extending the activity to take place over multiple class sessions.

Instructor Preparation and Materials

To complete this role-playing activity with students, the instructor should be prepared to (a) provide students with a brief reading or overview of the state of climate change communication, particularly in the USA (see below and Electronic Supplementary Materials, (ESM)-A, for instructor resources), (b) guide an open-ended discussion on the challenges of climate change communication, (c) facilitate the role-playing, and (d) guide an open-ended "debriefing" discussion after the activity.

This activity is most successful with students who have a basic knowledge of climate change, including an understanding of the central scientific concepts, the causes, and the present and future impacts. It is also useful if students have some familiarity with various types of solutions to climate change (e.g., knowing the difference between mitigation and adaptation). If climate change is a new topic for the class, the instructor can assign as an introduction to the issue either a documentary film (e.g., Orlowski 2013) or an overview chapter from one of the many available climate change handbooks (e.g., Henson 2011). In addition to this general background, students should prepare for this activity by reading one or more sources specifically about climate change communication (e.g., CRED 2009; see also ESM-A for other readings).

The materials needed to complete this activity are minimal. The instructor should create note cards or slips of paper (one for each student) on which are written "role descriptions" for the activity, for example, an oil rig worker, a climate scientist, a right-wing congressman, or a citizen of the Maldives. Ideally the total number of roles should equal the total number of students (so that each student has a different role). If the class is large, it is fine to have multiple students assigned to the same role (allowing collaboration on their research as desired). Fleshed out examples of these and other roles are listed in ESM-B. It can be useful if the instructor tailors the roles to highlight various topics already covered in the course. For example, if the students have learned about the Alberta tar sands, then one role could be a tar sands engineer and another role could be a member of a local Mikisew Cree First Nation community. The instructor can similarly customize role descriptions to fit with the interests and demographics of the particular group of students or with issues specific to the school or local community.

For a basic version of this activity, in which students have conversations in pairs, the role description cards should be marked as either "group 1" or "group 2," with the two groups corresponding to roles that are more on the "climate change skeptic" end of the spectrum or more on the "climate change believer" end of the spectrum. The students don't need to be told at the beginning of the activity what their group numbers correspond to; the numbers simply serve to place students into pairs during the in-class conversations. For ease of facilitation, there should be an equal number of roles in group 1 and group 2. If the class has an odd number of students, the instructor should be prepared to step into a role from either group as needed. Box 39.2 suggests other variations for implementing the activity.

Students can pick their roles out of a container (e.g., box, hat), or the instructor can randomly assign them. To participate in the conversations fully, students should complete background research about their assigned characters either before class or during the class meeting (see details below). If the latter, students will need electronic devices with Internet access in the classroom (provided or their own). Additionally, access to a chalk or whiteboard (or other writing surface) is useful for the debriefing discussions at the end of the activity.

Activities

Depending on course context, the instructor can begin as needed by fostering student reflection and leading a discussion (10–15 min) about the barriers to effective climate change communication and conversations. It is useful to start by asking: how many of you have been having, or have ever had, conversations about climate change outside of school? In the author's experience, most students answer "no." For students who answer "yes," it can be useful to query them about what factors made those conversations possible. The instructor can then ask students to complete a brief "think-pair-share" brainstorming activity in response to the question: why is climate change so difficult to talk about?

These opening questions provide a foundation for a large group discussion about any climate change communication readings that may have been assigned before class (see above and ESM-A). Students should be asked to share what they think are the key roadblocks/difficulties to effective climate change communication. Some of the common talking points are politics, lifestyle, religion, media bias, disinformation campaigns, the scale of the problem, feelings such as guilt or fear, and misunderstandings about the science, among others. It is less important that students identify all of these factors and, more so, that they are thinking generally about why climate change is a challenging issue for effective communication. The instructor can keep track of student responses on the board.

Next, the instructor should explain the learning goals and purpose of the role-playing activity (see section "Introduction" and Box 39.1). Providing this kind of framing for students helps prevent the activity from seeming artificial or falling flat. Next, if not already completed for homework, every student in the class should be given a "role card" and then conduct research (20–30 min) to develop their characters for the role-playing.

The instructor should provide guidelines for this process, including suggestions about how the students can fully flesh out the life stories of their characters and think about those characters' views on climate change. For instance, the instructor could encourage the student whose role is that of a parent worried about his/her children's future to come up with a name for the parent character and names for the children, to think about important events in the parent's life, and to consider what the parent's values and goals are. To relate this to climate change, the student might even conduct a quick Internet search for parents' organizations focused on solving climate change. More broadly, the instructor can encourage students to think about the following questions when doing research for their backstory: What are the desires, goals, and deeply held values of your character? How would these likely influence what your character thinks and feels about climate change? To provide additional guidance, the instructor can require that students fill out a "build a character" worksheet with more specific questions (see ESM-C).

When students have finished preparing their roles, the instructor announces that each student should pair up with someone from a different numbered group than his or her own (in the basic version of the activity, students with group 1 roles pair with students with group 2 roles) and find a spot in the room to sit and have a conversation about climate change (5–10 min). The instructor should explain the three requirements for the conversation: students must (1) stay in their roles for the duration of the role-play activity, (2) treat each other with respect, and (3) approach the conversation as if they genuinely want to learn about their partners' perspectives and share their own.

The instructor should monitor the conversations, providing reminders to help keep conversations on track as needed. However, even the conversations that seem to be "failing" are useful for student learning, especially when there is adequate time to discuss the reasons for such real or perceived failure during the debriefing. If time permits, the instructor can ask students to engage in additional conversations (see Box 39.2).

The debriefing stage of the activity is crucial. The instructor begins this portion of the activity by asking students to write a brief reflection about the process (~5 min), using one or more of the following prompts: What did you find most difficult about the conversation activity and why? Did you experience points of tension during your conversation, and if so, what were they? Were you able to resolve the tension or move past it? If you did not experience much tension in your conversation, why do you think that was the case? What did you and your conversation partner agree on or find common ground about? What did you learn from this activity?

The instructor should then lead a class discussion to elicit students' reflections, guiding summary and synthesis, and possibly recording key points on the board (20+ min, though this can carry over into the next class period). The key goal is to help students "talk about talking about" climate change. Most likely, some students will have experienced points of tension during their conversation (though they may have steered the conversation away from such tension). Thus, during this debriefing part of the activity, the instructor should guide students toward thinking about what those points of tension were, what the underlying differences were between their characters that created those tensions, and what strategies—including rhetorical choices—might have addressed those tensions (Nisbet 2009). In other words, how

> ## Box 39.2. Variations of Role-Play Conversations
>
> Theses variations work particularly well in a longer class period or with more advanced students:
>
> - After the initial conversations run their course, have students switch partners and engage in additional conversations. These conversations could be with someone with a different or similar position/outlook (i.e., different or same group number) on climate change. Engaging in additional conversations provides students with extra opportunities to learn about different perspectives on climate change and to hone their character acting.
> - Have students converse in groups comprised of three or more students, with roles organized into six categories corresponding to the multiple US climate change publics identified in the *Global Warming's Six Americas* study: alarmed, concerned, cautious, disengaged, doubtful, and dismissive (Leiserowitz et al. 2012). During debriefing, ask what it was like with more people included in the conversation. Was it easier or more challenging? Did the conversations go in different directions when there was a majority of one view on climate change in the group? Did one or two people dominate the discussion?
> - In groups of three or more, ask one of the students to be a neutral (perhaps silent) facilitator, mediator, or observer who records what happens and guides the conversation as needed. This person could also be tasked with helping the other participants find points of agreement and then reporting the outcomes to the class during the debriefing.

were the students able (or not) to find common ground in talking about climate change? To conclude with larger context, the instructor can ask students to generate a list of the reasons for why people's views about climate change differ (e.g., worldview, political affiliation, religion, age, profession) (Hulme 2009).

Follow-Up Engagement

- These questions can be used for extended discussion or writing assignments:
  - What makes it difficult to have conversations about climate change with people who have different worldviews and life stories?
  - What are some strategies for productively and respectfully talking about climate change when the person you are talking to does not share your beliefs, interests, and values?
  - Have you had climate change conversations or similar experiences outside of class, such as with family members or friends? How were those conversations similar or different from the role-play activity?

   – How might you approach such conversations in the future to engage more
     effectively with people who hold different views on climate change?

- Throughout the remainder of the course, ask students to have conversations
  with people outside of the class—friends, family, or community members—
  about climate change (or other course topics) using some of the communication strategies identified from the role-playing activity. These conversations
  can help students recognize that they have transferable skills for communicating about climate change with others. To facilitate learning, debrief about these
  experiences as a class.
- Ask students to work in groups or as a class to prepare their own guides or mani-
  festos about effective climate change communication. For instance, students
  might identify strategies such as "be aware of the person's job," "demonstrate
  your own understanding of climate change but don't act like a know-it-all," or
  "discuss the local impacts of climate change." Students can then compare their
  guide to those written by others (see ESM-A).

Connections

- When discussing other environmental sustainability-related issues, ask students to consider how their characters from this role-playing exercise would
  view them.
- When assigning students writing or presentation projects, ask them to consider
  how they will use different rhetorical strategies to appeal to readers or audience
  members who may not share their own views.
- Issues that relate to this chapter are discussed in other chapters of this volume
  including worldviews (Chap. 4), leadership skills (Chap. 7), navigating diverse
  priorities (Chap. 19), and effective communication about complex issues
  (Chaps. 37, 38, and 40).

References

Orlowski J (2013) Chasing ice. DVD. New Video Group, New York
Bliuc A, Mcgarty C, Thomas E et al (2015) Public division about climate change rooted in conflicting
    socio-political identities. Nat Clim Change 5:226–229
Center for Research on Environmental Decisions (CRED) (2009) The principles of climate change
    communication in brief. The Trustees of Columbia University. http://guide.cred.columbia.edu/
    guide/principles.html. Accessed 29 June 2015
Ding D, Maibach E, Zhao X et al (2011) Support for climate policy and societal action are linked
    to perception about scientific agreement. Nat Clim Change 1:462–566
Feldman L, Myers T, Hmielowski J et al (2014) The mutual reinforcement of media selectivity and
    effects: testing the reinforcing spirals framework in the context of global warming. J Commun
    64:590–611

Henson R (2011) The rough guide to climate change, 3rd edn. Rough Guides, London

Hobson K, Niemeyer S (2012) "What sceptics believe": the effects of information and deliberation on climate change scepticism. Publ Understand Sci 22:396–412

Hulme M (2009) Why we disagree about climate change: understanding controversy, inaction and opportunity. Cambridge University Press, Cambridge

Leiserowitz A, Maibach E, Roser-Renouf C (2012) Global warming's six Americas in September 2012. Yale Project on Climate Change Communication, New Haven

Marshall G (2014) Don't even think about it: why our brains are wired to ignore climate change. Bloomsbury, New York

Moser S, Dilling L (2011) Communicating climate change: closing the science-action gap. In: Dryzek J, Norgaard R, Schlosberg D (eds) The Oxford handbook of climate change and society. Oxford University Press, New York, pp 161–174

Nisbet M (2009) Communicating climate change: why frames matter for public engagement. Environment 51:514–518

Chapter 40
Writing Letters to the Editor to Promote Environmental Citizenship and Improve Student Writing

Andrew J. Schneller

Introduction

Resolving environmental and sustainability issues is complex and fraught with debate. Civil, community discourse that includes a myriad of stakeholder perspectives is necessary to educate and involve the public (e.g., see Chaps. 19 and 31). Successful engagement requires that citizens be able to understand, assess, and successfully communicate their own and others' perspectives about complex issues across scales (Jacobson 2009; Pezzulo 2010; Chaps. 29 and 39). To help students improve these skills, the process of reading and writing "letters to the editor" (LTE) provides an experiential and authentic assignment that also fosters learning about timely environmental problems and politics (Wermuth 2006). LTE writing can be "situated as [an] actual activity in which [students] participate in real-world settings" (Doyle 2000, p. 1) thus engaging them as environmental citizens in their communities. This chapter describes how reading and writing LTEs can be used to help students develop and articulate their personal perspectives about environmental and sustainability issues and improve their writing skills.

A benefit of using LTEs as pedagogy is that they can easily be submitted for publication, which provides an external audience and added motivation to enhance student engagement. Because the LTE section is one of the most widely read parts of newspapers and news websites (Jacobson 2009; Sierra Club North Star 2014), students who write LTEs that are subsequently published can reach a wider audience

Electronic supplementary materials: The online version of this chapter (doi:10.1007/978-3-319-28543-6_40) contains supplementary material, which is available to authorized users.

A.J. Schneller, Ph.D. (✉)
Environmental Studies Program, Skidmore College, Saratoga Springs, NY, USA
e-mail: aschnell@skidmore.edu

with their environmental perspectives. The assignment can also be empowering for students who may not have many opportunities to have their voices heard in a public forum. In the author's experience, 40–50 % of submitted student LTEs per class have been accepted for publication (see two examples in Electronic Supplementary Materials (ESM-A)). After 4 years of facilitating this activity, it is evident (through written student evaluations) that students (and instructors) appreciate how reading and writing LTEs helps them acquire useful skills while simultaneously connecting them to their communities through environmental advocacy.

Learning Outcomes

After completing this activity, students should be able to:

- Analyze and critique the framing of complex public policy issues/problems within popular media.
- Frame and write succinct and persuasive arguments about their environmental perspectives for a broad public audience of community members (e.g., businesses, elected officials, and agencies at multiple levels of government).
- Engage in public environmental discourse with confidence and civility.
- Recognize the value and role of engaging in public environmental discourse for communicating their environmental perspectives.

Course Context

- Developed for an upper-level environmental policy course for environmental studies majors with 15–30 students
- Time needed to complete the full activity is two 80-min classes but adjustments can be made for one class and shorter periods
- Adaptable to courses that include discussions of US public lands and oceans policies, environmental communication, environmental governance and management, or other sustainable development and sustainability issues

Instructor Preparation and Materials

This activity is taught in two parts and requires computers for student use in the classroom (if available). If computers are unavailable in the classroom, it is suggested that instructors reserve a computer lab (e.g., in the library). For parts 1 and 2, instructors should provide a handout (LTE guide) to students that describes the reasoning and logistical process of writing letters to the editor. The most useful resource for this process has been created free for public use by national chapters of the Sierra Club (ESM-B). Instructors should edit this information as needed for their course.

For part 1, instructors should choose a collection of opinion-editorial (op-ed) essays on one topic for the students to read before class that they will write about in a practice letter. Although this activity could be used with many topics, it was designed to coincide with the discussion of current media events and environmental policy decisions. The chosen op-eds should provide students with a broad enough spectrum of public perspectives so as to better understand the diversity of stakeholder interests in the issue. A set of three example op-eds about salvage logging on Bureau of Land Management lands in southern Oregon is provided in ESM-C. This sequence of op-eds was carefully selected for advancing student analytical and writing goals, as they represent varied legal, anthropocentric, and biocentric values of community stakeholders, thus highlighting the complexity and diversity of arguments within the environmental decision-making arena.

For a briefer implementation, durations for part 1 suggested below can be reduced and/or step D (letter writing) can be completed as homework before the in-class peer review. Alternatively, it is possible to skip the practice LTE assignment altogether and begin with part 2. (In the author's experience, the longer and shorter approaches work, but the in-class writing and peer review process provides more continuity between writing and editing and receiving immediate feedback for improvement.)

Activities

Part 1: The Practice Letter to the Editor

(a) As homework, hand out the chosen op-eds (e.g., ESM-C) that students should read before the letter-writing class. Students should be instructed to highlight the strongest arguments in each of them. Tell students that in the next class they will be required to take a stance on the issues discussed in the op-eds and that they will be writing a letter to the editor that is informed by these pieces.

(b) On the day of the exercise, hand out the "letter to the editor guide" (Sierra Club 2014, ESM-B). Discuss the benefits of LTEs as mentioned in the Sierra Club guide, and discuss the main points for successfully publishing a LTE (10 min). Consider inviting a journalism professor, an environmental journalist, or an editor (if available in the community or via Skype) as a guest speaker who can provide an insider's perspective about the publishing industry.

(c) Instruct students to take on a fictional community role for their practice LTE (for the logging case study, e.g., a bird watcher, hiker, logger, Earth First!er, mountain biker, ecologist, fisher, wilderness advocate, botanist, motocross fanatic, government resource management agent, etc.). This community stake-holder's perspective will be used for writing their practice LTE. Alternatively, the instructor may choose to assign students their roles randomly.

(d) In class, using a computer, students will individually write a ~250-word LTE (30 min). Explain that students who are having difficulty writing their LTEs will soon have assistance from a peer editor.

(e) When they are finished writing their LTE, students should choose or be assigned a partner for peer editing. Suggest to students that this step is helpful because chances are, if the peer editor does not understand something in the LTE, the editor of the newspaper will also be at a loss.

(f) The peer review process (30 min) involves students sharing their practice LTEs with their partner (e.g., via email, Google Docs, or switching computers). Instructors should explain that peer reviewers should write comments related to grammar, content, organization, and any additional information that is needed to clarify and/or improve the main points and arguments in the LTE. While preparing their comments, reviewers should follow the instructions in the LTE guide (ESM-B). After both students finish writing their comments, each should, in turn, provide them to the author and verbally summarize their comments and suggested edits. Note that it may be helpful for students to edit their partner's LTEs if they assume the role of a newspaper editor or member of the general public. Instructors may also choose to request that peer reviewer comments are turned in for evaluation.

(g) Students should make edits to their LTE in class or at home. Instructors may wish to collect these for evaluation based, in part, on common issues that will arise such as exceeding the 250-word limit; failure to include a witty title; failure to mention the author, date, and title of an op-ed or article to which the LTE is referring; failure to explicitly state the desired outcomes; failure to use multiple short paragraphs as opposed to one giant block paragraph; failure to stick to one or two main points; and lack of new/unique compelling arguments in relation to their chosen or assigned stakeholder role.

(h) As appropriate, the instructor can facilitate a 10-min whole-class discussion to debrief about the class' opinions (e.g., preferred outcomes for the proposed salvage logging sale on public lands; see ESM-C).

Part 2: Letters to the Editor for Submission to a Newspaper

Building on part 1, part 2 asks students to write an LTE for submission to an actual media outlet, most likely a newspaper or possibly a weekly or bimonthly magazine. Part 2 should commence in the class immediately following part 1.

(a) Students should be asked to browse newspapers (online, 30 min) from their hometown or the community in which the university is located to find a recent article or op-ed related to the class material (they should have been published within the last 2 weeks to increase the probability of students' letters being published). This can be an article that students feel strongly about or that contains content for which they're curious or even unfamiliar. The importance of finding a "hometown" or university community article is somewhat important, as newspapers are more likely to publish letters from their local readership/residents. However, students in the author's classes have also had success publishing nationally regarding more general domestic and international environmental issues (e.g., climate change, endangered species, and public lands). Instructors should explain to students that they will have a greater likelihood of publication success if they choose a smaller newspaper, instead of a high-profile national paper.

(b) Students should be instructed to find the LTE submission process for the publication of their selected article and note the word limit (usually 150–400 words).

(c) Once students have read their article or op-ed, they should be given 30 min for the LTE writing process in class.

(d) After the letter writing, they will again work for 20 min with a peer editor to improve their LTEs (see step (f), part 1 above). If some students are working more slowly and do not have time to both read the newspaper article and also write their LTE, students can work at their own pace and finish the process outside of class.

(e) When students have finished their edits in class (if time permits) or later at home, they should be asked to submit their LTE online to the publication and their instructor. To confirm submission to the publication, students should either carbon copy the instructor on an email submission or take a "screenshot" of the completed Web page submission form (or confirmation page) where they have submitted their LTE. As noted in the Sierra Club guidelines (ESM-B), students can also be invited to send the letter to their local elected or agency officials to increase the impact of this assignment and their citizenship engagement.

(f) Instructors should ensure sufficient time for debriefing about the value of this lesson, highlighting the extent to which students learned to critically read about, reflect on, and assess complex environmental issues and their ability to express themselves concisely in writing with conviction and clarity. Discussion can also focus on the value of LTEs in relation to the potential broader impacts of contributing to the public discourse (e.g., what difference does it make if people write LTEs and to what extent can anyone's mind be changed by a letter?).

Follow-Up Engagement

• After students submit their LTEs to a newspaper, they should look online for successful publication. Students should send the instructor online links to the LTE publications. Instructors and students can then discuss any public replies/comments that have been posted online in relation to the student LTEs. Further, environmental issues that have been resolved after publication of the LTE can be highlighted as part of a class discussion.

- One option for increasing the rigor of this assignment is for students to write an opinion editorial (op-ed) on an issue, typically about 800–1,000 words in length (see ESM-C). For this, students, individually or in groups of two or three (possibly in conjunction with a community stakeholder, NGO representative, or other "community influential"), contact an editor of a newspaper with an initial writing proposal (lede) to obtain the green light for formal submission and possible publication. Tips for writing good op-eds, hooks, and ledes can be found online from *The OpEd Project* (www.theopedproject.org/). Alternatively, a collective, whole-class op-ed piece can be written regarding one agreed-upon topic, with each student writing pieces that are integrated into one op-ed (by the instructor or as a class project, perhaps online in a shared document). (See Wermuth (2006) for an example of this process.)
- A further extension is for instructors to post students' LTEs and the original op-eds online in a classroom management platform and have students comment (civilly) on each other's ideas.

Connections

- Because students are writing persuasively about contemporary and pressing environmental issues and problems, there will be fruitful future ground (at other times in the course) for instructors to call upon students in class to discuss what they have written about in their LTEs. In the author's experience, students genuinely care about the complex issues they read about and become de facto experts on them and are thus more likely to contribute and participate in class discussions and debates.
- If the LTE and op-ed writing process is a student's first foray into writing for environmental advocacy, they will acquire a better understanding of the challenges and complexity of the environmental communications and policy arenas. A desired longer-term outcome of this heightened understanding (and the writing process) is that the acquired skill set and experience will provide a foundation for future community engagement efforts (e.g., leadership roles; see Chap. 7).

References

Doyle W (2000) Authenticity. Paper presented at the annual meeting of the American Educational Research Association, New Orleans, LA, April 2000

Jacobson SJ (2009) Communication skills for conservation professionals. Island Press, Washington, DC

Merriam-Webster Dictionary (2015) Op-ed. http://www.merriam-webster.com/dictionary/op-ed. Accessed 29 June 2015

Pezzulo PC (2010) Teaching environmental communication through rhetorical controversy. In: Reynolds HL, Brondizio ES, Robinson JM (eds) Teaching environmental literacy: across campus and across the curriculum. Indiana University Press, Bloomington, pp 98–107

Sierra Club North Star Chapter (2014) Writing a letter to the editor. http://northstar.sierraclub.org/priorities/at_the_capitol/write_a_letter_to_the_editor/tips_for_a_good_letter/. Accessed 29 June 2015

The OpEd Project (2014) http://www.theopedproject.org/. Accessed 29 June 2015

Wermuth L (2006) Debating issues through opinion-editorials and letters to the editor. In: Perry JL, Jones SG (eds) Quick hits for educating citizens. Indiana University Press, Bloomington, pp 28–30

Chapter 41
Captioning Political Cartoons from Different Perspectives as a Tool for Student Reflection

Katharine A. Owens

Introduction

Political cartoons use humor and thought-provoking satire to reveal public perceptions about controversial topics (Hirt 2011); they can also soothe or inflame the mood around issues. They often allow for social and political agenda setting, reflect the value of issues in society, and seek to influence thought on a given subject (Sani et al. 2012). In analyzing political cartoons on climate change, researchers have noted that cartoons provide commentary on social power structures and public knowledge as well as give voice to alternative messages (Manzo 2012, Eide 2012). In these ways, political cartoons can serve as a rich tool to expose students to the nuances and depth of social discourse around environmental issues.

Political cartoons are underutilized in classroom settings even though they can be quite effective in teaching theoretical and conceptual material and improving critical thinking skills (Hammett and Mather 2011). In a study measuring students' environmental concept knowledge when taught with and without political cartoons, Toledo et al. (2013) found that the use of cartoons resulted in significantly higher learning gains. Examining political cartoons helps students hone their analytical skills while enlivening class discussions, in part, by sparking enthusiasm (Doherty 2002). Further, Bickford (2012) emphasizes that creating political cartoons fostered higher levels of critical thinking in students; he found

Electronic supplementary materials: The online version of this chapter (doi:10.1007/978-3-319-28543-6_41) contains supplementary material, which is available to authorized users.

K.A. Owens (✉)
Politics and Government and Environmental Studies, University of Hartford,
Hartford, CT, USA
e-mail: kowens@hartford.edu

that when used as a complement to or in lieu of writing assignments, political cartooning allowed students to effectively communicate abstract ideas to their peers. The goal of this chapter's learning activity is to cultivate students' critical thinking, analytical thinking, and writing skills by creating and discussing political cartoon captions from different perspectives and writing a short reflective piece on the exercise.

Learning Outcomes

After completing this activity, students should be able to:

- Consider discourse on an environmental issue by creating succinct political cartoon captions.
- Express varied perspectives about an environmental issue by writing cartoon captions from different viewpoints.
- Interpret the discourse revealed by cartoons through discussion.

Course Context

- Developed for a public policy course for political science majors with 20–30 students and has been used in undergraduate environmental policy and introductory environmental studies courses with 10–15 students
- 20 min in one class meeting
- Students should be familiar with the focal theme or topic in order to communicate multiple perspectives
- Adaptable to courses of any class size and level that include discussions of environmental issues and can be modified for longer durations by allowing more time for captioning and discussion or by adding extensions as described in the "Follow-Up Engagement" section

Instructor Preparation and Materials

To prepare, the instructor should access blank political cartoons or remove text from captioned cartoons. The Association of American Editorial Cartoonists (AAEC) maintains cartoons specifically for classroom use on their website, including environmental ones (see Electronic Supplementary Materials (ESM)-A). Cartoons published in papers can be collected and adapted for classroom use by removing their captions (which is condoned by the AAEC). Additional cartoon sources can be found in ESM-A.

When choosing political cartoons, it is helpful to consider the advice of Doherty (2002), who notes the importance of selecting suitable cartoons, cautioning that overly obvious or complicated cartoons may not lend themselves to classroom use. As she suggested, "you should have a clear purpose in mind for what you want the cartoon(s) to accomplish. Simply because a cartoon is about (a given issue) does not mean that the issues it raises are the ones you want to discuss" (Doherty 2002, p. 259).

The instructor should select at least two cartoons on a theme of interest. The instructor will make copies of the cartoons (to be handed out in class), giving one cartoon to each student group (of two to four students). In a very large class, several cartoons on the same theme may be used to allow for variation. There can certainly be overlap; it is not necessary that each group has a different cartoon. The total number of cartoons may ultimately be determined by how many the instructor can find on the theme. The cartoons' subject(s) should align with topics covered/discussed in the course to date because deeper knowledge of an issue provides students with the vocabulary necessary to consider more than one perspective (i.e., not simply their own opinions). The instructor should digitally scan the cartoon image(s) (all of those to be handed out to students) to create a slideshow in which each cartoon is projected as it is discussed. It may also be helpful to collect a few published political cartoons, both humorous and thought provoking, to include in a presentation to introduce students to the medium.

To become acquainted with discourse about images, some of the following articles may be helpful. A clear, short introduction to the use of environmental political cartoons in the classroom can be found in Hirt (2011). For additional perspectives on political cartoon analysis, Conners' (2007) article is concise and accessible, albeit focused on presidential candidates. O'Neill and Nicholson-Cole (2009) describe research on the effectiveness (or lack thereof) of fear-based climate-change imagery, enabling consideration of the broader implications of the messages in political cartoons. While longer, Dunaway's (2008) article may be best for sharing with students as it explores visual imagery (photographs and cartoons) and environmental discourse associated with the first Earth Day to illuminate the framing of this event.

Activities

Introduction (2–4 min)

Tell students that their task is to create a caption for the given political cartoon. Let them know that it can be funny or serious, but should remain connected to the issue or class topic. Take heed of this advice for writing captions from The New Yorker's Robert Mankoff (2010): "See, you can't come up with a funny caption if you don't come up with any captions at all. Therefore, the first thing to do is to lower the bar. Lower it to the ground, so you can step over it and generate as many potential captions as possible. You can always raise it later. The easiest way to do this is by using the words that immediately pop into your head as a jumping-off point for your captions"

(p. 1). In other words, brainstorming is the most effective way to generate potential captions. Encourage students to suggest many different captions and perhaps select their favorite at the end. This allows creativity to flow within the group, giving students the freedom to offer ideas without pressure.

Captioning (6–10 min)

Ask students to form small groups. Give each group a copy of the blank cartoon and have students spend 3–5 min creating a caption for the cartoon. Instructors should walk around to the groups to provide encouragement and feedback (remind students to add captions freely and edit later) and help students stay on task. Urge students to try captions that might inspire people to think about the issue in a new way. Ask students to imagine different audiences with questions such as: What caption would you write if this were to be submitted to the state's largest newspaper or to your school newspaper? What if you were writing for your parents' friends or your roommates?

After brainstorming their initial caption, ask students to then create a caption from an alternate perspective (i.e., from an environmentalist and then a business owner, from a conservative and then a liberal, from an older and then a younger person) for the same cartoon (in an additional 5–7 min).

Discussion (6–12 min)

The instructor projects each cartoon, asking student groups to share their best two captions, each one reflecting a different perspective. Then, as time permits, further discussion about each cartoon can be fostered using questions suggested by Doherty (2002, p. 259), including: "Who is in the cartoon? What is the context of the cartoon? Why is it funny? How are the various characters drawn and what clues does this offer us about the cartoonist's viewpoint? If characters are portrayed through visual metaphors, what does the metaphor tell us?"

The instructor should also use pointed questions to help students analyze different positions and perspectives on the issue raised by a cartoon, such as: Who are the stakeholders? What are their stakes? How might different stakeholders think differently about this issue? The instructor can challenge groups to provide alternative perspectives from a range of stakeholder groups. Can the stakeholders be arranged in coalitions? What might be the stakes of each coalition? What might be examples of acceptable or unacceptable policies for each group? This process can be repeated for each cartoon. Further examples of discussion questions can be found in ESM-B. To conclude the discussion, the instructor should encourage students to reflect on the way we as a society communicate about important environmental issues and how cartoons can help and/or hinder effective discourse.

Follow-Up Engagement

- After the discussion, students can be given a reflective short essay assignment to allow them to thoughtfully reflect on the activity along with a rubric for its assessment (available in ESM-B).
- For extended discussion or as prompts for other assignments, the following questions can be used:
  - How can messaging improve or obfuscate understanding about an environmental issue?
  - Are all perspectives given equal time on environmental issues? Should they be given equal time?
  - Can messaging be used to build understanding rather than highlight difference?
  - Juxtapose and compare cartoons from different publications, authors, and even countries. Is there a difference in messages?
- Have students work in small groups or alone to create their own cartoons on a different topic (skill in drawing is not required, stick figures can be quite effective). Bickford (2012) provides a guide for leading students through this process, including the use of concept mapping and student examples.

Connections

- Writing a succinct message relates to creating public relations campaigns for sustainability-related community engagement (e.g., effective letters to the editor; see Chap. 40, this volume).
- Detecting political and other biases can help students understand and critique the rhetorical messages in various media (e.g., Chaps. 35, 36, 42and 43, this volume).
- Thinking about the way various stakeholders consider an issue can provide insight into political decision-making and how to engage in more successful dialogue (e.g., Chaps. 19 and 39, this volume). This relates to issues of environmental justice (see Chaps. 27–30, this volume).

References

Bickford J (2012) Original political cartoon methodology and adaptations. Soc Stud Res Pract 7:91–101

Conners J (2007) Popular culture in political cartoons: analyzing cartoonist approaches. PS Polit Sci Polit 40:261–265

Doherty B (2002) Comic relief: using political cartoons in the classroom. Int Stud Perspect 3:258–270

Dunaway F (2008) Gas masks, Pogo, and the ecological Indian. Am Q 60:67–99

Eide E (2012) Visualizing a global crisis. Constructing climate, future and present. Confl Commun Online 11:1–16

Hammett D, Mather C (2011) Beyond decoding: political cartoons in the classroom. J Geogr High Educ 35:103–119

Hirt P (2011) Teaching U.S. history with environmentally themed cartoons. OAH Mag Hist 25:45–48

Mankoff R (2010) How to win the cartoon caption contest. The New Yorker. http://www.newyorker.com/cartoons/bob-mankoff/how-to-win-the-cartoon-caption-contest. Accessed 7 May 2015.

Manzo K (2012) Earthworks: the geopolitical visions of climate change cartoons. Polit Geogr 31:481–494

O'Neill S, Nicholson-Cole S (2009) Fear won't do it: promoting positive engagement with climate change through visual and iconic representations. Sci Commun 30:355–379

Sani I, Abdullah M, Abdullah F, Ali Z (2012) Political cartoons as a vehicle of setting social agenda: the newspaper example. Asian Soc Sci 8:156–164

Toledo A, Yangco T, Espinosa A (2013) Media cartoons: effects on concept understanding in environmental education. World J Environ Res 3:13–32

Chapter 42
Analyzing Nature as a Persuasive Tool in Advertisements

Rebecca Duncan

Introduction

Manufacturers of consumer goods and advertising professionals who promote them have a long, complex relationship with nature and perceived human connections with the environment (Howlett and Ragion 1992). Nature—ranging from stunning landscapes to adorable and fear-inducing wild animals—has served as a persuasive strategy since the inception of print advertising. When displayed beside or behind a product, a visual portrayal of nature and the "common desires, beliefs, and values" it invokes can convince us to open our wallets (Sturgeon 2009, p. 13). More generally, corporations use natural imagery and ideologies to suggest a core commitment to sustainability and environmental stewardship (Carlson et al. 1993; also see Chap. 23, this volume).

Analyzing the role of nature in advertising can position students for further inquiry in three academic contexts. First, in lessons about writing or persuasion, such an activity provides practice in analyzing rhetorical strategies, particularly *logos*, *pathos*, and *ethos*, using familiar products and images. Second, in the context of environmental literature, philosophy, or ethics, the activity can establish a critical framework with which to approach complex readings and contemplations on human relationships with the natural world. Third, for general sustainability studies, this work provides a glimpse into the messages and practices of self-identified "green" companies and products, introducing the distinction between a fundamental and enacted commitment versus attempts to "greenwash," or project a false image of a

Electronic supplementary materials: The online version of this chapter (doi:10.1007/978-3-319-28543-6_42) contains supplementary material, which is available to authorized users.

R. Duncan (✉)
English Department, Meredith College, Raleigh, NC, USA
e-mail: duncanr@meredith.edu

© Springer International Publishing Switzerland 2016
L.B. Byrne (ed.), *Learner-Centered Teaching Activities for Environmental and Sustainability Studies*, DOI 10.1007/978-3-319-28543-6_42

product's and its producer's environmental responsibility (Underwriters Laboratory 2015). To these ends, the activity described in this chapter involves a close reading of references to nature in advertisements, with the broader goal of helping students practice and improve analytical and sustainability thinking skills.

Learning Outcomes

After completing this activity, students should be able to:

- Articulate a human/nature relationship expressed or implied in an ad.
- Identify argumentative strategies at work in an ad (e.g., in terms of *logos, ethos, pathos,* or other paradigm).
- Assess the effectiveness of an ad for influencing an audience of their peers.

Course Context

- Developed for an introductory course in persuasive writing with 20–30 students
- 50 min in one class meeting
- Students can be asked to gather advertisements but no other preparation or background is needed
- Adaptable to any courses that explore human relationships with nature or "green" advertising and its potential abuses

Instructor Preparation and Materials

Preparation will vary, depending on which of the three scenarios (described above) fits the course: writing and persuasion, human/nature relationships, or sustainability practices. An overview of relevant scholarship for each approach is provided in Electronic Supplementary Materials (ESM)-A. Those who use this activity to teach argument should present an overview of the Aristotelian concepts of *logos, ethos,* and *pathos* in advance. *Logos* refers to a logical argument; *ethos* reflects ethics, social responsibility, character, and values; and *pathos* is an emotional appeal. Axelrod and Cooper (2014) provide basic definitions, and elaboration of these concepts in relation to nature-based ads is in ESM-A.

To ensure a range of examples with clear associations between a product and nature, it is recommended that the instructor gather a set of sample advertisements. Images of plants and flowers, animals, landscapes, geographical features, and water are typical associations with the natural world featured in ads. Although trends shift over time, products typically promoted using nature include soaps and shampoos, makeup, automobiles, fashion, hunting equipment, vacations, and a variety of foods.

Nature-related ads can be gathered from magazines, newspapers, direct mailings, and other sources. *Smithsonian* and *Real Simple* are fruitful sources, as are business and women's magazines. A handful of excellent examples is provided in Sturgeon (2009). Online sources are also readily available. In addition to online versions of commercial and trade magazines, resources can be located with the search terms "environmental advertising examples," "nature advertising examples," and "animal advertising." The websites of product manufacturers and advertising agencies also feature sample ads. A sample ad and analysis appear in ESM-B and ESM-C.

In addition to the instructor-provided ads, or alternately as a pre-class assignment, students can be asked to find sample ads and bring them to class. If so, they may need help distinguishing between advertising and editorial content; this confusion is the only difficulty that has arisen in several years of conducting this activity by the author.

Activities

The steps below reflect the most basic iteration of the activity as it might be used in an introductory lesson. The analyses can take on greater complexity in a course in which students have read additional source material on environmental advertising and greenwashing and have the requisite product knowledge to assess the advertising claims.

1. Introduce the idea that we can deduce views about human relationships with nature by examining the use of nature in advertising of consumer products (5–7 min).
2. Give each student a sample ad that depicts nature in relation to a product. Ask them to work in groups of two to four to examine the visual presentation of nature in each other's samples and respond to the following questions, reproduced on the printable worksheet in ESM-D (3 min of individual analysis, 5 min for each student to share with group members, 5 min to prepare key points for presentation).

   (a) What images are used? What colors and actions dominate the ad? How are people reacting to the scene depicted?
   (b) Next, examine the explicit and implied messages. Does the ad's copy (the headline and additional writing) mention the natural world? What is said? What additional messages are implied or suggested by the text and/or images?
   (c) Together, how do the message and visual elements attempt to persuade the audience (e.g., through logos, ethos, and/or pathos)? Explain.
   (d) Does the ad seem to mislead the reader in any way about the product's association with the natural world? Is there any other kind of contradiction

between depictions of nature or sustainability and the product's function as
you understand it?

(e) What does this ad suggest about sustainability and human relationships with
nature?

3. Ask several of the groups to share one of their ads and observations. Engage the
class in finding patterns across the ads and contradictions within them (3–5 min
for each presentation and class comments).

4. As a class, compile a list of human/nature relationships suggested by the ads (for
examples, see ESM-E—10 min).

5. As students present their observations, the instructor should, as desired, draw
attention to the use of logos, ethos, and pathos and address any misunderstandings of these concepts. Connections can also be made between the human/nature
relationships that students discover in the ads and other course topics and
readings.

Follow-Up Engagement

- The list of human/nature relationships generated by the class can serve as a
framework for further analysis. Students can be asked to locate additional examples of nature in ads and write informal journal entries or formal essays in which
they analyze several ads in light of these relationships (and logos/ethos/pathos)
or they can apply the framework to a literary work about the environment.
- Students can practice rhetorical strategies by creating ads of their own that use
images of and written references to nature.

Connections

- This activity, when supplemented by the cited readings, can prepare students to
address environmental and sustainability issues related to corporate social
responsibility, consumer behavior, or behavioral economics (see Chaps. 22 and
23, this volume).
- Business-focused lessons addressing a particular product or economic sector
might include further study on greenwashing, especially if students have or can
acquire the necessary product knowledge to contrast advertising claims with a
product's performance or characteristics.
- This activity can be used to connect sustainability and environmental issues with
lessons about media studies, cross-cultural, or global communication.

Acknowledgements This work has been supported by the Environmental and Sustainability
Initiative of Meredith College, Raleigh, North Carolina, which is funded by a grant from the
Margaret A. Cargill Foundation.

References

Axelrod RB, Cooper CR, Warriner AM (2014) Reading critically/writing well: a reader and guide. Bedford/St. Martin's, Boston, MA

Carlson L, Grove SJ, Kangun N (1993) A content analysis of environmental advertising claims: a matrix method approach. J Advert 22:27–39

Howlett M, Ragion R (1992) Constructing the environmental spectacle: green advertisements and the greening of the corporate image, 1910–1990. Environ Hist Rev 16:53–68

Sturgeon N (2009) Environmentalism in popular culture: gender, race, sexuality, and the politics of the natural. University of Arizona Press, Tucson, AZ

Underwriters Laboratory (2015): The Sins of Greenwashing: home and family edition. http://sin-sofgreenwashing.com/. Accessed 21 Aug 2015

Chapter 43
Making and Assessing Art in the Sustainability Classroom

Bruno Borsari

Introduction

Education for sustainability calls for students to become active agents of change and good stewards of the Earth. One way to foster these outcomes is through the process of creating art. Arts and the larger culture are inextricably linked, and consequently, educating for sustainability by employing the arts may help foster a culture of sustainability (Lehtonen et al. 2013). Art and the art-making process have potential for helping students restore their connections to the environment. Scientific information can inspire artworks (Borsari et al. 2008), and participation in art making engages learners emotionally and cognitively (Emery 2013). The arts can move the head and heart toward a clearer understanding of the world in which we live (Borsari and De Grazia 2015). Further, art making has potential for sparking imagination which enhances creativity, yielding an engaging learning environment. Thus, an instruction milieu that mimics an art studio can be more compelling and help improve sustainability education (Jacobson et al. 2007).

The main focus of the art-making activity described below is fostering students' positive expressions about sustainable change—rather than producing "good art"—while enhancing their understanding about, and experience of, the aspects of "art thinking" that differ from other modes of thinking, such as scientific. In this activity, students are invited to make art that is inspired by sustainability issues. It embraces

Electronic supplementary materials: The online version of this chapter (doi:10.1007/978-3-319-28543-6_43) contains supplementary material, which is available to authorized users.

B. Borsari (✉)
Department of Biology, Winona State University, Winona, MN, USA
e-mail: BBorsari@winona.edu

a systemic philosophy which applies directly to a transformative and community-learning pedagogy (Clover 2000). In sum, this activity aims at connecting sustainability across disciplines while helping all students think creatively when challenged by the complex themes or concepts taught in class.

Learning Outcomes

After completing this activity, students should be able to:

- Express their feelings about sustainability through creative, collaborative artwork, without being inhibited by perceived inabilities to produce "good art".
- Communicate to classmates how their artwork conveys a sustainability message.
- More critically evaluate sustainability messages conveyed in artworks and transfer the acquired skills to all art.

Course Context

- Developed for a large (>150 students), intro nonmajors biology course
- Used after an introductory class on sustainability, in which a clickers' case about the ecological footprint concept had been used (Borsari 2009)
- Can be completed in a 50-min class period and is adaptable to any class size

Instructor's Preparation and Materials

Before students begin the activity, the instructor should be prepared to give a short, introductory presentation (3 min) to explain the work to be done and show examples of artworks from students (see Electronic Supplementary Materials (ESM)-A) and professional artists (painters, poets, musicians, photographers, and others). For example, a short video of Joan Baez singing "Where have all the flowers gone" (https://www.youtube.com/watch?v=0LZ2R2zW2Yc) demonstrates an especially good artwork which blends poetry and music. Sources for and examples of environmental and sustainability-themed art are provided in ESM-B. For projection of introductory slides and websites as needed, ensure that the classroom is equipped with a projector and Internet access.

The instructor should also prepare to lead a short discussion (2–3 min) by asking a few probing questions (suggested below). These are derived from current literature about arts and the environment (Curtis et al. 2014; Borsari and De Grazia 2015) that the instructor can consult in preparation for the art-making class. Additional background readings are suggested in ESM-C.

The instructor should make copies of the Artwork Assessment Tool (AWAT) for the evaluation of the students' art pieces (ESM-D). Each student will need one copy for each artwork to be reviewed; preparing ten forms per student should suffice although the number will depend on the number of students and groups in the class and time allocated for each student to evaluate each artwork (see below for context to help guide decisions). For the art-making work, instructors may wish to gather materials to provide (markers, pencils, etc.). In the author's experience, art can be successfully made on computers so students should be told to bring those if desired. In addition, engaging art professors in various departments on campus (e.g., music, graphic design, arts, dance, and theater) by inviting them to partici-pate in the art-making class is encouraged to demonstrate multidisciplinary edu-cation pedagogy and to inspire these colleagues to offer this or similar activities to their classes as well.

In an attempt to reduce students' inhibition to art making, instructors could share an art piece they have created (see ESM-B for an example from the author) and utilize a theme to contextualize the art making. In the author's experience, a theme for the activity can strengthen the learning outcomes of this activity more success-fully by focusing students' attention, allowing them to be more productive in a short time frame. For example, "The ecological footprint dilemma" case proposed by Borsari (2009) has been used to prepare students for the art-making class; as a "warm-up" lesson used in the class meeting before the art making one, it enabled students to reflect and discuss sustainability more critically through a relevant story which applies to their lives.

Activities

This activity is implemented in six parts: (1) a mini-presentation by the instructor, (2) a discussion about critical questions, (3) a brainstorming session to select a sus-tainability concept, (4) art making, (5) a gallery walk, and (6) wrap-up session.

(1) and (2): The instructor should organize students into groups of three to five individuals after providing a short introductory presentation (3 min; see above). Then, students can be asked to comment on and discuss questions (6 min) such as: What are the differences (or similarities) in the thinking process applicable to arts and/or science making? What is good art? What is the worth of art making in mod-ern society? How does art relate to sustainability? The instructor should allow some time (2 min) to reporting responses to the whole class by soliciting groups to share their answer.

(3): Then, facilitate a 3-min brainstorming session to list (on a white/chalkboard), as a class, a dozen or so concepts that relate to sustainability that could be used as themes for art projects (Box 43.1) (or that relate to an umbrella theme, e.g., the eco-logical footprint). Each group is asked to select a concept that will be used to inspire an original art piece of a chosen medium (e.g., poem, song, visual art, or dance). Students should be asked to identify an objective for the art piece they are going to create. More specifically, each group will need to clearly articulate the message, idea, and/or emotion that they want to communicate through their artwork.

Box 43.1. Sustainability Issues and Themes That Can Translate Well into Student Art

- Genetically modified organisms (GMOs) and their implication for environmental and public health
- Conservation of pollinator species, habitat, and biodiversity
- Oil and food production as human population continues to grow
- Soil and water: Are they renewable or nonrenewable resources?
- Personal reflections on one's ecological footprint (see Borsari 2009)

(4): At this point, the instructor should ask students to make their art piece with the following guidelines:

- They are *not* compelled to produce "good art" but rather do their best to identify a way to express their chosen sustainability concept in an artistic piece.
- They should have fun participating in this class activity.
- They must give a title to their artwork.
- They must indicate the sustainability concept conveyed by the art (and including it in the title is suggested).
- They will have 15 min to complete their artwork. (Although it may seem that this is insufficient time, in the author's experience, the time allocated functions well to ensure that students keep focused on the activity.)
- The instructor will keep the time and ask students to stop working on their art piece when time is up.
- When done, students will share their art with their classmates in a gallery walk when students will be asked to evaluate each other's artwork.
- As desired, students can be asked to send photos or the original files (for digital work) to the instructor by email for archiving and, if appropriate, instructor assessment.

(5): At least 20 min should be devoted to the "gallery walk" when the art pieces are displayed and assessed by students. (To display digital works, students' laptop computers can be employed.) Students' assessment of the art is a very important component of this class activity and it can vary by class size. In a large class, the instructor will ask students to arrange themselves around the room to present their works. Each group will designate a presenter who will stand next to the art piece to briefly share the title of, conceptual inspiration for, and the message that the authors want to convey through the work. (Presentation time (e.g., 50–60 s) should be set based on class size and duration.) Non-presenting students serve as evaluators who will move from artwork to artwork, listen to the presentations, and complete an AWAT form (ESM-D) for each; if time permits, the recommended minimum is ten evaluations per student. In the author's experience,

AWAT forms can be completed in about 10 s. Evaluators can assess each artwork once and they cannot assess their own group's piece. The instructor should tell students when to move to the next presentation. So that every student can be an evaluator, presenters should switch their role with another group member after every five presentations.

In smaller classes, the art pieces can be presented one by one to the whole class (if needed for digital pieces, through a projector). The instructor will invite each group to the front of the classroom to share their work in 1–2 min (or longer as time permits). All group members or a spokesperson should briefly share the title, the sustainability concept that inspired the artwork, and the message that the authors want to convey. At the end of each presentation, students in the audience will fill out the AWAT form while the professor invites another group to present.

(6): During the last three (or more) minutes of class, the instructor wraps up the exercise by facilitating a discussion to solicit students' reflections. These questions could be used as prompts (see more prompts in ESM-E):

- What was learned in this activity?
- How did the artistic process (from the selection of the concept to art making) work for the groups?
- What were the challenges (if any) in selecting the sustainability concept?
- Was the activity enjoyable or frustrating?
- How does art making connect to the course?

Follow-Up Engagement

- The instructor can analyze the AWAT data by calculating averages for each category for each artwork and, at the following class meeting, announce the class' most successful works.
- Students can be assigned to write a reflection paper that aims to verify the learning outcomes of the art-making exercise. (See ESM-E for a suggested assignment sheet and rubric.)
- Collaborate with art directors to set up future students' displays of artworks related to sustainability at galleries on or near to campus (e.g., see Curtis (2011) for an example with extension workers).

Making Connections

Connections to the art-making activity could be made across a course and curriculum by:

- Discussing contrasting "ways of knowing," as when the concept of "science" is introduced (e.g., see Chaps. 9 and 14, in this volume).

- Exploring how emotions and attitudes expressed through the arts are important to consider as they relate to people's behaviors (i.e., discussing issues of environmental psychology; see Chaps. 4, 6 and 22, in this volume).
- Analyzing how the arts and other imagery can be used to engage people in thinking about public policy and social problems that relate to environmental challenges (also see Chaps. 41 and 42, in this volume).

References

Borsari B, De Grazia E (2015) Bruno's garden. North Dakota Quarterly 8:64–84

Borsari B (2009) The ecological footprint dilemma. In: National Center for Case Study Teaching in Science. University of Buffalo NY. http://sciencecases.lib.buffalo.edu/cs/collection/detail.asp?case_id=543&id=543 Accessed 18 Jan 2015

Borsari B, Richardson R, Scott Plummer A et al (2008) Earth's Precambrian era as a common evolutionary theme in two art courses at Winona State University. J Evol Outreach Educ 1(2):172–178

Clover DE (2000) Community arts as environmental education and activism: a labour and environment case study. Convergence 33(4):19–30

Curtis DJ (2011) Using the arts to raise awareness and communicate environmental information in the extension context. J Agr Educ Ext 17(2):181–194

Curtis DJ, Reid N, Reeve I (2014) Towards ecological sustainability: observations on the role of the arts. S A P I E N S 7(1):1–15

Emery S (2013) Making the case for "Arts for Sustainability": A study of educators' views of education for sustainability. Master's Thesis, University of Tasmania at Launceston, Australia

Jacobson SK, McDuff MD, Monroe MC (2007) Promoting conservation through the arts: outreach for hearts and minds. Conserv Biol 21:7–10

Lehtonen K, Juvonen A, Ruismäki H (2013) The multiple aims of arts education to support sustainable development. In: Härkönen U (ed) Proceedings of the 10th international JTEFS/BBCC conference on sustainable development, culture, education, reports and studies in education, humanities, and theology No. 7, University of Eastern Finland, Joensuu, Finland, p 263–273

Index

A

Advertisement, 307–310
Advertising, 36, 174, 178, 307–310
Affective, 2, 5, 7, 8, 10, 43, 73, 75, 166, 168–170
Agriculture, 80, 85, 95, 103, 138, 151, 165, 166, 169, 209, 213
Allee effects, 130
Amenity, 212
Apollo 13, 90–94
Art, 270, 295–297, 313–318
Attitude, 1, 5, 7, 8, 43, 48, 73, 103, 167, 169, 218, 318

B

Behavior
    change, 57, 63, 64, 166, 169, 173
    pro-environmental, 57
Belief, 7, 8, 18, 19, 26, 44–46, 48, 49, 69, 150, 153, 261, 285, 286, 290, 307
Bias, 85, 261, 262, 268, 270, 271, 275, 277, 288
Biodiversity, 4, 55, 75, 78, 95, 98, 100, 111, 112, 114, 126, 137–141, 144, 147, 165, 185, 199, 232, 253, 255–258, 266, 271, 277, 316
Biome, 102
Business, 4, 34, 58, 132, 175, 181, 185, 274, 294, 304, 309, 310

C

Carbon
    cycling, 37, 42
    dioxide (CO_2), 81, 83, 179, 184, 188–191, 199, 248, 256, 258
    footprint, 177, 178, 205
Cartoon (political), 301, 305
Cell phone, 33, 93
Citizenship, 46, 205, 293–298
City, 51, 227–232, 240, 257
Civic engagement, 64
Civil discourse, 293, 294
Clean Air Act, 95, 239, 240, 243, 247
Climate, 51, 89, 92, 97, 101, 126, 158, 160, 165, 249, 257, 287
    change, 37, 42, 43, 55, 63, 67, 73, 75, 78, 94, 100–102, 109, 126, 179, 191, 193, 196, 209, 213, 219, 232, 258, 266, 268, 271, 277, 285–291, 297, 301
Coal, 40, 184, 196, 199, 201, 203, 213, 239
Communication, 14, 49, 63, 129, 238, 273, 276, 279, 281, 283, 285–288, 291, 294, 298, 310
Community engagement, 298, 305
Compassion, 43
Competition, 54, 94, 106–108, 118, 123, 131, 269
Complexity, 36, 65, 70, 89, 102, 130, 152, 201–205, 227–228, 276, 281, 295, 298, 309
Concept map, 37, 39, 217, 218, 305
Conflict, 35, 43, 129, 146, 192, 237, 285
Consensus (scientific), 80, 123
Conservation, 76, 95, 106, 111, 120, 121, 124, 127, 130, 134, 137, 139, 140, 143–148, 151, 185, 199, 202, 258, 271, 277
Consumer goods, 307
Consumerism, 35, 174, 175, 266
Corporate social responsibility, 36, 184, 310
Corporation, 182, 199, 307
Cradle-to-cradle design, 42
Cultures, 5, 43, 99, 313

© Springer International Publishing Switzerland 2016
L.B. Byrne (ed.), *Learner-Centered Teaching Activities for Environmental and Sustainability Studies*, DOI 10.1007/978-3-319-28543-6

Printed in the United States
By Bookmasters